Lecture Notes in Mathematics Vol. 1205

ISBN 978-3-540-16784-6 © Springer-Verlag Berlin Heidelberg

B.Z. Moroz

Analytic arithmetic in algebraic number fields

Errata

p.2, last line:	write "$w_p(\mathfrak{A}) = 0$" instead of "$w_p(\mathfrak{A}) = 1$"
p.5, line 2:	write "$\lambda \in \hat{X}^*$" instead of "$\lambda \in X^*$"
p.13, formula (6):	write "$< \chi \vert \psi >= 0$" instead of "$< \chi \vert \psi >= 1$"
p.35, formula (14):	write "$\tilde{S}_1$" instead of "S_1"
p.38, formula (30):	write "$(\frac{\vert 1+s\vert}{2\pi})^{n/2}$" instead of "$\frac{\vert 1+s\vert}{2\pi})^{n/2}$"
p.43, formula (14) and line 8:	write "$\alpha \leq \dots$" instead of "$\alpha \geq \dots$"
p.46, line 6:	write "$O(nx \exp(\dots))$" instead of "$O(x \exp(\dots))$"
p.46, line 12:	write "$\psi(t) = c_4(\log \varphi(t,\chi))^{-1}$" instead of "$\psi(t) = c_4 \log \varphi(t,\chi)$"
p.47, formula (38):	write "$\frac{\log T}{2\psi(t)}$" instead of "$\frac{\log T}{\psi(t)}$"
p.48, line 1:	write "$O(nx \exp(\dots))$" instead of "$O(x \exp(\dots))$"
p.48, formula (41):	write "$O(x \sum_{j=1}^{m} n_j \exp(\dots))$" instead of "$O(x \sum_{j=1}^{m} \exp(\dots))$"
p.54:	Lemma 2 holds only under the additional assumption that $f(\alpha) \neq 0$ and $f(s) \neq 0$ when $\vert s\vert = 9/4$
p.56, formula (20):	write "$\frac{g(\chi)}{s-1}$" instead of "$g(\chi)(\frac{1}{s} + \frac{1}{s-1})$"
p.56, last line:	write "... ≤ 2, except possibly at $s = 0$, the" instead of "... ≤ 2, the"
p.57, formula (22):	write "$R(s) = \frac{g(\chi)(r_1+r_2)}{s} + O\dots$" instead of "$R(s) = O\dots$"
p.58, formula (27):	write "$\dots g(\chi)x - \dots$" instead of "$\dots g(\chi)x + \dots$"
p.60, line 11:	write "5) μ $(\dots$" instead of "5) $(\dots$"
p.67, formula (20):	write "$\omega(k)$" instead of "$w(k)$"
p.83, formula (29):	write "$\vec{\chi}$" instead of "χ" (twice)
p.84, formula (38):	write "$S_0(\vec{\chi})$" instead of "$S_0(\chi)$"

p.89:	Theorem 1 is incorrect as it stands; it should be replaced by [103, Proposition 1]
p.100:	Lemma 2 is incorrect as it stands; it should be replaced by [103, Lemma 5]
p.103:	now a sharper version of Lemma 4 is available, see [103, Lemma 4]
p.111, last line:	write "$d_3 = 1$" instead of "$d_3 = 2$"
p.127, formula (10):	write "$\sum_{\mathfrak{p}\in S_0(k),\lvert\mathfrak{p}\rvert=p}$" instead of "$\mathfrak{p} \in S_0(k), \lvert\mathfrak{p}\rvert = p$"
p.128, line 8:	write "$E_{\mathfrak{A}}$" instead of "E"
p.141, line 1:	write "norm-form" instead of "non-form"
p.148, formula (37):	write "$\omega(k)$" instead of "$x(k)$"
p.149, formula (38):	write "$\geq$" instead of "$>$"
p.152, formula (4):	write "$C_1(\vec{k})$" instead of "$C_1(k)$"
p.164, line 4:	write "$\neq \emptyset$" instead of "$\neq O$"
p.164, formula (5):	write "$C_2(\vec{k}) > 0$" instead of "$_2(\vec{k}) > 0$"
p.167, line 9:	write "$a_j \in \mathbb{R}^{d_j}$" instead of "$\lvert a_j\rvert \in \mathbb{R}^{d_j}$"
p.168, line 4:	write "these authors, see also the two survey articles [105], [106]." instead of "these authors."
p.168, line 3 from below:	write "[53], [101]." instead of "[53]."
p.169, line 9:	write "[66] (cf. also [103])." instead of "[66]."
p.169, last line:	write "theorem II.5.1 (cf. [102])." instead of "theorem III.5.1."
p.170, line 13:	write "[98]. In the mean time Conjecture 1 has been proved [104]." instead of "[98]."
p.174, reference [65]:	write "24 (1986), pp. 272-283" instead of "to appear"
p.174, reference [68]:	write "Israel Journal of Mathematics, 60 (1987), pp. 1-21." instead of "M.P.I. für Mathematik Preprint 86-14, Bonn, 1986."
p.176, reference [100]:	write "On the number of integral points on a norm-form variety in a cube-like domain, Journal of Number Theory, 27 (1987), pp. 106-110." instead of "A footnote to my recent paper (submitted for publication)."

p.176, add the following references:

[101] M.D. Coleman, A zero-free region for Hecke L-functions, Mathematica, 37 (1990), pp. 287-304.

[102] N. Kurokawa, On the meromorphy of Euler products, I, II, Proceedings of the London Mathematical Society, 53 (1988), pp. 1-47, 209-236.

[103] B.Z. Moroz, On a class of Dirichlet series associated to the ring of representations of a Weil group, Proceedings of the London Mathematical Society, 56 (1988), pp. 209-228.
[104] B.Z. Moroz, Equidistribution of Frobenius classes and the volumes of tubes, Acta Arithmetica, 51 (1988), p. 269-276.
[105] B.Z. Moroz, Scalar product of Hecke L-functions and its application, Advanced Studies in pure maths, 21 (1992), pp. 153-171.
[106] B.Z. Moroz, Scalar products of Dirichlet series and the distribution of integer points on toric varieties, Zapiski Nauchnykh Seminarov POMI, 322 (2005), pp. 135-148.

Lecture Notes in Mathematics

Edited by A. Dold and B. Eckmann

Subseries: Mathematisches Institut der Universität und Max-Planck-Institut
für Mathematik, Bonn – vol. 7

Adviser: F. Hirzebruch

1205

B. Z. Moroz

Analytic Arithmetic in Algebraic Number Fields

Springer-Verlag

Berlin Heidelberg New York London Paris Tokyo

Author

B. Z. Moroz
Max-Planck-Institut für Mathematik, Universität Bonn
Gottfried-Claren-Str. 26, 5300 Bonn 3, Federal Republic of Germany

Mathematics Subject Classification (1980): 11D57, 11R39, 11R42, 11R44, 11R45, 22C05

ISBN 3-540-16784-6 Springer-Verlag Berlin Heidelberg New York
ISBN 0-387-16784-6 Springer-Verlag New York Berlin Heidelberg

Library of Congress Cataloging-in-Publication Data. Moroz, B. Z. Analytic arithmetic in algebraic number fields. (Lecture notes in mathematics; 1205) "Subseries: Mathematisches Institut der Universität und Max-Planck-Institut für Mathematik, Bonn – vol. 7." Bibliography: p. Includes index. 1. Algebraic number theory. I. Title. II. Series: Lecture notes in mathematics (Springer-Verlag; 1205. QA3.L28 no. 1205 [QA247] 510 [512'.74] 86-20335
ISBN 0-387-16784-6 (U.S.)

Printing and binding: Beltz Offsetdruck, Hemsbach/Bergstr.
2146/3140-543210

Introduction.

This book is an improved version of our memoir that appeared in Bonner Mathematische Schriften, [64]. Its purpose is twofold: first, we give a complete relatively self-contained proof of the theorem concerning analytic continuation and natural boundary of Euler products (sketched in Chapter III of [64]) and describe applications of Dirichlet series represented by Euler products under consideration; secondly, we review in detail classical methods of analytic number theory in fields of algebraic numbers. Our presentation of these methods (see Chapter I) has been most influenced by the work of E. Landau, [40], [42], E. Hecke, [24], and A. Weil, [91] (cf. also [87]). In Chapter II we develop formalism of Euler products generated by polynomials whose coefficients lie in the ring of virtual characters of the (absolute) Weil group of a number field and apply it to study scalar products of Artin-Weil L-functions. This leads, in particular, to a solution of a long-standing problem concerning analytic behaviour of the scalar products, or convolutions, of L-functions Hecke "mit Grössencharakteren" (cf. [63] for the history of this problem; one may regard this note as a résumé of Chapter II, if you like). Chapter III describes applications of those scalar products to the problem of asymptotic distribution of integral and prime ideals having equal norms and to a classical problem about distribution of integral points on a variety defined by a system of norm-forms. Chapter IV is designed to relate the contents of the book to the work of other authors and to acknowledge our indebtedness to these authors.

I should like to record here my sincere gratitude to Professor P. Deligne whose remarks and encouragement helped me to complete this work. This book, as well as [64], has been written in the quiet atmosphere of the Max-Planck-Institut für Mathematik (Bonn). We are grateful to the Director of the Institute Professor F. Hirzebruch for his hospitality and support of our work. The author acknowledges the hospitality of the Mathematisches Institut Universität Zürich, where parts of the manuscript have been prepared.

Bonn-am-Rhein, im März 1986.

Table of contents

Notations and conventions.

We shall use the following notations and abbreviations:

$\emptyset$	empty set
$:=$	"is defined as"
$A\setminus B$	the set theoretic difference
$\mathbb{N}$	the set of natural numbers (including zero)
$\mathbb{Z}$	the ring of natural integers
$\mathbb{Q}$	the field of rational numbers
$\mathbb{R}$	the field of real numbers
$\mathbb{R}_+$	the set of positive real numbers
$\mathbb{C}$	the field of complex numbers
A^*	the group of invertible elements in a ring A
$\hat{G}$	the set of all the simple (continuous) characters of a (topological) group G
$\hat{k}$	a fixed algebraic closure of the field k
1	denotes the unit element in any of the multiplicative groups to be considered
$\{x\mid P(x)\}$	is the set of objects x satisfying the property $P(x)$

card S, or simply $|S|$, stands for the cardinality of a finite set S;

$E\mid F$	is an extension of number fields: $E \supseteq F$
$[E:F]$	denotes the degree of $E\mid F$
$G(E\mid F)$	denotes the Galois group of a finite extension $E\mid F$
(α)	is a principal ideal generated by α
$\mathfrak{a}\mid\mathfrak{b}$	means divisor $\mathfrak{a}$ divides $\mathfrak{b}$
$\lvert\mathfrak{a}\rvert$	is the absolute norm, that is $N_{E/\mathbb{Q}}\mathfrak{a}$, of a divisor $\mathfrak{a}$ in a number field E
$\lvert x\rvert$	is the absolute value of a complex number x
$\vec{\mathfrak{a}},\vec{\chi},\vec{k}$	stand for finite sequences (of a fixed length) of divisors, characters, fields, etc.
$\mathrm{Im}\,\varphi$	is the image of the map φ

Ker φ	is the kernel of the homomorphism φ
Re s	is the real part of s in $\mathbb{C}$
Im s	is the imaginary part of s in $\mathbb{C}$
$f\circ g$	denotes the composition of two maps, so that $(f\circ g)(a) = f(g(a))$
$\Gamma(s)$	is the Euler's gamma-function
l.c.m.	is the least common multiple
g.c.d.	is the greatest common divisor
$\rho\|_H$	denotes the restriction of a map ρ to the set H
G^c	denotes sometimes (the closure of) the commutator subgroup of a (topological) group G
$A \otimes B$	is the tensor product of A and B
$A \oplus B$	is the direct sum of A and B

References of the form: theorem I.2.3, lemma 1.1, proposition 2, corollary I.A2.1 mean theorem 3 in §2 of Chapter I, lemma 1 in §1 of the same chapter, proposition 2 in the same paragraph and corollary 1 in Appendix 2 of Chapter I, respectively; the same system is used for references to numbered formulae. Relations proved under the assumption that Riemann Hypothesis, Artin-Weil conjecture or Lindelöf Hypothesis are valid shall be marked by the letters R, AW, L, respectively, before their number.

Every paragraph is regarded as a distinct unit, a brief relatively self-contained article; thus we try to be consistent in our notations throughout a paragraph but not necessarily over the whole chapter. In the first three chapters we avoid bibliographical and historical references which are collected in the Chapter IV.

Chapter I. Classical background.

§1. On the multidimensional arithmetic in the sense of E. Hecke.

Let k be an algebraic number field of degree $n = [k:\mathbb{Q}]$. Consider the following objects:

v is the ring of integers of k;

S_1 and S_2 are the sets of real and complex places of k respectively, $S_\infty = S_1 \cup S_2$;

S_o is the set of prime divisors of k identified with the set of non-archimedean valuations;

$S := S_o \cup S_\infty$ is the set of all primes in k;

$r_j := |S_j|$, $j = 1,2$, so that $n = r_1 + 2r_2$;

k_p is the completion of k at p for $p \in S$;

U_p is the group of units of k_p for $p \in S_o$;

w_p is the valuation function on k_p normalised by the condition $w_p(k_p^*) = w_p(k^*) = \mathbb{Z}$, $p \in S_o$;

$I_o(k)$ is the monoid of integral ideals of k;

$I(k)$ is the group of fractional ideals of k;

$(\alpha) = \prod_{p \in S_o} p^{w_p(\alpha)}$ is the principal ideal generated by α in k^*, we extend the valuation function w_p to I and write $\mathfrak{a} = \prod_{p \in S_o} p^{w_p(\mathfrak{a})}$ for $\mathfrak{a} \in I$;

J_k is the idèle group of k;

$C_k := J_k/k^*$ is the idèle-class group of k, where k^* is embedded diagonally in J_k;

$X := \prod_{p \in S_\infty} k_p$ is regarded as a n-dimensional $\mathbb{R}$-algebra. The group of units v^* acts freely as a discrete group of transformations on the multiplicative group X^*, the action being given by

$$x \mapsto \varepsilon x, \; x \in X^*, \; \varepsilon \in v^*,$$

where k is embedded diagonally in X. Obviously,

$$X^* \cong (\mathbb{Z}/2\mathbb{Z})^{r_1} \times T^{r_2} \times \mathbb{R}_+^{r_1+r_2} , \tag{1}$$

where $T = \{\exp(2\pi i\varphi) \mid 0 \leq \varphi < 1\}$ denotes the unit circle in $\mathbb{C}^*$. Let m be the order of the maximal finite subgroup of k^*; by a theorem of Dirichlet,

$$v^* \cong \mathbb{Z}^{r_1+r_2-1} \times \mathbb{Z}/m\mathbb{Z} . \tag{2}$$

One can show that, in accordance with (1) and (2),

$$X^*/v^* \cong (\mathbb{Z}/2\mathbb{Z})^{r_o} \times \mathbb{R}_+ \times \mathcal{T} , \tag{3}$$

where $r_o \leq \max\{0, r_1-1\}$ and $\mathcal{T}$ is a real $(n-1)$-dimensional torus. The diagonal embedding of k into X gives rise to a monomorphism

$$f_o \colon k^*/v^* \to X^*/v^*$$

of the group of principal ideals of k into the group (3). Let $g\colon X^*/v^* \to \mathcal{T}$ denote the natural projection map of X^*/v^* on the torus $\mathcal{T}$. The composition of these maps $g \circ f_o$ can be continued to a homomorphism

$$f\colon I(k) \to \mathcal{T} , \quad f_{|P_k} = g \circ f_o , \tag{4}$$

where $P_k := k^*/v^*$. Let $\mathfrak{m} \in I_o(k)$ and let $\mathfrak{m}_\infty \subseteq S_1$. One defines a subgroup

$$I(\mathfrak{m}) = \{\mathfrak{a} \mid \mathfrak{a} \in I(k),\ w_p(\mathfrak{a}) = 1 \quad \text{for } p|\mathfrak{m},\ p \in S_o\}$$

of $I(k)$ and a subgroup

$$P(\tilde{\mathfrak{m}}) = \{(\alpha) \mid \alpha \in k^*, \quad \alpha \equiv 1(\mathfrak{m}), \quad \sigma_p(\alpha) > 0 \quad \text{for} \quad p \in \mathfrak{m}_\infty\}$$

of P_k, where σ_p denotes the natural embedding of k in k_p for $p \in S$, and $\tilde{\mathfrak{m}} := (\mathfrak{m}, \mathfrak{m}_\infty)$. The ray class group

$$H(\tilde{\mathfrak{m}}) := I(\mathfrak{m})/P(\tilde{\mathfrak{m}})$$

is a finite group of order

$$|H(\tilde{\mathfrak{m}})| = h\varphi(\tilde{\mathfrak{m}}),$$

where

$$h = |H(1,\emptyset)| , \quad H(1,\emptyset) = I(k)/P_k$$

are the class number and the class group of k respectively, and

$$\varphi(\tilde{\mathfrak{m}}) := \text{card } (I(\mathfrak{m}) \cap P_k)/P(\tilde{\mathfrak{m}}).$$

For a smooth subset τ of $\mathfrak{T}$ and a ray class A in $H(\tilde{\mathfrak{m}})$ one defines two functions

$$\iota(\cdot;A,\tau) : \mathbb{R}_+ \to \mathbb{N} \quad ,$$
$$\pi(\cdot;A,\tau) : \mathbb{R}_+ \to \mathbb{N} \quad ,$$

by letting

$$\iota(x;A,\tau) = \text{card } \{\mathfrak{a} \mid \mathfrak{a} \in A \cap I_o, \quad f(\mathfrak{a}) \in \tau, |\mathfrak{a}| < x\}$$

and

$$\pi(x;A,\tau) = \text{card}\,\{p \mid p \in A \cap S_o,\ f(p) \in \tau,\ |p| < x\}.$$

We are interested in obtaining asymptotic estimates for $\iota(x;A,\tau)$ and $\pi(x;A,\tau)$ as $x \to \infty$. To this end one defines grossencharacters and studies L-functions associated with these characters.

A <u>grossencharacter modulo</u> $\tilde{\mathfrak{m}} = (\mathfrak{m}, \mathfrak{m}_\infty)$ is, by definition, a character χ of $I(\mathfrak{m})$ for which there is λ in $\hat{X}^*$ such that

$$\chi((\alpha)) = \lambda(\alpha) \quad \text{whenever} \quad \alpha \in k^*, \quad (\alpha) \in P(\tilde{\mathfrak{m}}). \tag{5}$$

Let $\tilde{\mathfrak{m}}_i = (\mathfrak{m}_i, \mathfrak{m}_{\infty i})$ and let χ_i be a grossencharacter modulo $\tilde{\mathfrak{m}}_i$, $i = 1,2$. If $\mathfrak{m}_1 | \mathfrak{m}_2$, $\mathfrak{m}_{\infty 1} \leq \mathfrak{m}_{\infty 2}$, and $\chi_1(\mathfrak{a}) = \chi_2(\mathfrak{a})$ for $\mathfrak{a} \in I(\mathfrak{m}_2)$, we write $\chi_1 \leq \chi_2$. A grossencharacter χ is called a <u>proper grossencharacter</u> if $\chi_1 \leq \chi$ implies $\chi_1 = \chi$. Given a proper grossencharacter χ modulo $\tilde{\mathfrak{m}} = (\mathfrak{m}, \mathfrak{m}_\infty)$ we call $\tilde{\mathfrak{m}}$ the <u>conductor</u> of χ and write

$$\tilde{\mathfrak{m}} = \widetilde{\mathfrak{f}(\chi)}, \quad \mathfrak{m} = \mathfrak{f}(\chi), \quad \mathfrak{m}_\infty = \mathfrak{f}_\infty(\chi).$$

One continues a proper grossencharacter χ to a multiplicative function $\chi\colon I_o(k) \cup I(\mathfrak{f}(\chi)) \to T \cup \{0\}$ by letting $\chi(\mathfrak{a}) = 0$ for $\mathfrak{a} \in I_o(k) \setminus I(\mathfrak{f}(\chi))$. To simplify our notations we write $\alpha \equiv 1(\tilde{\mathfrak{m}})$ for $\alpha \in k^*$ whenever $\alpha \equiv 1(\mathfrak{m})$, $\sigma_p(\alpha) > 0$ for $p \in \mathfrak{m}_\infty$. Let $v^*(\tilde{\mathfrak{m}}) = \{\varepsilon \mid \varepsilon \equiv 1(\tilde{\mathfrak{m}})\}$; it is a subgroup (of finite index) of v^* regarded as a transformation group of X^*. We embed $\mathbb{R}_+$ diagonally in X^*: $t \mapsto (t^{1/n},\dots,t^{1/n})$, $t \in \mathbb{R}_+$, $t^{1/n} \in \mathbb{R}_+$. The following result is a generalisation of (3).

<u>Lemma 1</u>. The character group

$$\hat{X}^*(\tilde{\mathfrak{m}}) = \{\lambda \mid \lambda \in X^*;\ \lambda(\varepsilon x) = \lambda(x) \text{ for } x \in X^*,\ \varepsilon \in v^*(\tilde{\mathfrak{m}});$$
$$\lambda(t) = 1 \quad \text{for} \quad t \in \mathbb{R}_+\}$$

is isomorphic to $(\mathbb{Z}/2\mathbb{Z})^{r_o} \times \mathbb{Z}^{n-1}$ with $r_o \leq r_1$.

<u>Proof</u>. For $\lambda \in \hat{X}^*$, $x \in X^*$ we have

$$\lambda(x) = \prod_{p \in S_\infty} |x_p|^{it_p} \left(\frac{x_p}{|x_p|}\right)^{a_p} \tag{6}$$

with $t_p \in \mathbb{R}_+$, $a_p \in \mathbb{Z}$; moreover $a_p \in \{0,1\}$ when $p \in S_1$. Here x_p denotes the p-component of x, so that $x_p \in k_p$. Condition $\lambda(t) = 1$ is equivalent to the equation

$$\sum_{p \in S_\infty} t_p = 0.$$

The second condition $\lambda(\varepsilon x) = \lambda(x)$ for $x \in X^*$, $\varepsilon \in v^*(\tilde{\mathfrak{m}})$ leads, in view of the Dirichlet theorem on units, to a system of linear equations for the exponents $\{t_p, a_p\}$. Solving these equations one finds a system of generators for $\hat{X}^*(\tilde{\mathfrak{m}})$. We refer for these calculations to [24] (cf. also [23], §9).

A grossencharacter χ satisfying (5) is said to be <u>normalised</u> if $\lambda \in \hat{X}^*(\tilde{\mathfrak{m}})$.

<u>Lemma 2</u>. For every λ in $\hat{X}^*(\tilde{\mathfrak{m}})$ there is a grossencharacter χ modulo $\mathfrak{m}$ such that $\chi((\alpha)) = \lambda(\alpha)$ whenever $\alpha \in k^*$ and $(\alpha) \in P(\tilde{\mathfrak{m}})$.

<u>Proof</u>. See [24] (cf. also [23], §9; [91]).
The following assertion can be easily deduced from Lemma 1 and Lemma 2.

<u>Proposition 1</u>. The group of normalised grossencharacters modulo $\tilde{\mathfrak{m}}$ is

isomorphic to

$$(\mathbb{Z}/2\mathbb{Z})^{r_o} \times \mathbb{Z}^{n-1} \times \hat{H}(\mathfrak{m}).$$

For $x \in k_p$ one writes

$$\|x\|_p = \begin{cases} |x| & \text{when } p \in S_1 \\ |x|^2 & \text{when } p \in S_2 \\ |p|^{-w_p(x)} & \text{when } p \in S_o \end{cases}$$

Let $x \in J_k$ and let x_p be the p-component of x, we set then $\|x\| = \prod_{p \in S} \|x_p\|_p$. By the product formula,

$$\|\alpha\| = 1 \quad \text{for} \quad \alpha \in k^*,$$

therefore the map $\alpha \mapsto \|\alpha\|$ is well defined on C_k. Let

$$C_k^1 = \{\alpha | \alpha \in C_k, \ \|\alpha\| = 1\}$$

be the subgroup of idèle-classes having unit volume. The group C_k^1 is known to be compact. The group X^* can be identified with the subgroup $\{x | x \in J_k, \ x_p = 1 \ \text{for} \ p \in S_o\}$ of J_k, so that $\mathbb{R}_+$ embedded diagonally in X^* may be regarded as a subgroup of C_k. It follows then that

$$C_k = \mathbb{R}_+ \times C_k^1. \tag{7}$$

There is a natural homomorphism id: $J_k \to I(k)$ of J_k on $I(k)$ given by the equation

$$\mathrm{id}\, x = \prod_{p \in S_o} p^{w_p(x_p)} \quad \text{for} \quad x \in J_k.$$

Let $\mu \in \hat{C}_k$, let $\mathfrak{f}(\mu)$ be the conductor of μ (defined as, e.g., in [93], p. 133) and let $\mathfrak{f}_\infty(\mu)$ be set of those primes in S_1 at which μ is ramified; write $\widetilde{\mathfrak{f}(\mu)} = \{\mathfrak{f}(\mu), \mathfrak{f}_\infty(\mu)\}$. One can define a character χ_μ on $I(\mathfrak{f}(\mu))$ by the equation

$$\chi_\mu(\mathfrak{a}) = \mu(x) \quad \text{for} \quad \mathfrak{a} \in I(\mathfrak{f}(\mu)), \quad x \in \mathrm{id}^{-1}(\mathfrak{a}),$$

if one regards μ as a character of J_k (trivial on k^*). It follows from definitions that χ_μ is well defined since μ is constant on $\mathrm{id}^{-1}(\mathfrak{a})$ for $\mathfrak{a} \in I(\mathfrak{f}(\mu))$.

Proposition 2. The function $\mathfrak{a} \mapsto \chi_\mu(\mathfrak{a})$ is a proper grossencharacter and $\widetilde{\mathfrak{f}(\chi_\mu)} = \widetilde{\mathfrak{f}(\mu)}$; it satisfies (5) with $\tilde{\mathfrak{m}} = \widetilde{\mathfrak{f}(\mu)}$ and λ equal to the restriction of μ to X^* (regarded as a subgroup of J_k), in particular, χ_μ is normalised if and only if $\mathbb{R}_+ \subseteq \mathrm{Ker}\,\mu$. If χ is a proper grossencharacter, there is one and only one μ in $\hat{C}_k$ such that $\chi = \chi_\mu$.

Proof. See [91], p. 9 - 10 (or [23], §9).
We denote the group of proper normalised grossencharacters by $gr(k)$ and remark that

$$gr(k) \cong \hat{C}_k^1.$$

Proposition 1 defines a fibration of $gr(k)$ over the set of (generalised) conductors. Let $\chi \in gr(k)$ and suppose that χ satisfies (5) with λ of the shape (6); we call a_p, t_p appearing in (6) exponents of χ and write $a_p = a_p(\chi)$, $t_p = t_p(\chi)$.

Let now

$$s_p(\chi)=\begin{cases} a_p(\chi)+it_p(\chi) & , \quad p\in S_1 \\ \frac{1}{2}(|a_p(\chi)|+it_p(\chi)), & p\in S_2 \end{cases},$$

and let

$$G_p(s) = \begin{cases} \pi^{-s/2}\Gamma(s/2), & p\in S_1 \\ (2\pi)^{1-s}\Gamma(s), & p\in S_2 \end{cases}.$$

For $s \in \mathbb{C}$, $\chi \in gr(k)$ one defines a Dirichlet series

$$L(s,\chi) = \sum_{n=1}^{\infty} c_n(\chi)n^{-s} \quad , \tag{8}$$

where

$$c_n(\chi) = \sum_{|\mathfrak{a}|=n} \chi(\mathfrak{a}), \qquad \mathfrak{a}\in I_o(k), \tag{9}$$

is a finite sum extended over the integral ideals of k whose norm is equal to n. The series (8) converges absolutely for $\mathrm{Re}\, s > 1$ and in this half-plane it can be decomposed in an Euler product:

$$L(s,\chi) = \prod_{p\in S_o} (1-\chi(p)|p|^{-s})^{-1}. \tag{10}$$

One extends (10) by adding the gamma-factors at infinite places:

$$\Lambda(s,\chi) = L(s,\chi) \prod_{p\in S_\infty} G_p(s+s_p). \tag{11}$$

By a theorem of E. Hecke, [24] (cf. also [93], VII §7) the function

$$s \mapsto \Lambda(s,\chi)$$

can be meromorphically continued to the whole complex plane $\mathbb{C}$ and satisfies a functional equation:

$$\Lambda(s,\chi) = W(\chi)a(\chi)^{\frac{1}{2}-s} \Lambda(1-s,\bar{\chi}), \qquad (12)$$

where $a(\chi) = |D|\cdot|\mathfrak{f}(\chi)|$, D denotes the discriminant of k and $|W(\chi)| = 1$. The function

$$s \mapsto L(s,\chi) - \frac{\omega(k)g(\chi)}{s-1} ,$$

where $g(\chi) = 0$ for $\chi \neq 1$ and $g(1) = 1$, is holomorphic in $\mathbb{C}$. The residue of $L(s,1)$ at $s = 1$ is given by the equation: $\omega(k) = 2^{r_1+r_2}\pi^{r_2}R\ h(m\sqrt{|D|})^{-1}$, where R is the regulator of k and m denotes the order of the group of roots of unity contained in k^*. We write, for brevity,

$$\zeta_k(s) = L(s,1),\ \zeta_{\mathbb{Q}}(s) = \zeta(s),$$

and let

$$L_\infty(s,\chi) = \prod_{p\in S_\infty} G_p(s+s_p(\chi)), \qquad (13)$$

so that

$$\Lambda(s,\chi) = L(s,\chi)L_\infty(s,\chi). \qquad (14)$$

§2. Group theoretic intermission.

Let G be a compact group and let μ be the Haar measure on G normalized by the condition $\mu(G) = 1$. In what follows we do not distinguish between equivalent representations and consider, as we may without loss of generality, only finite dimensional unitary representations. Let $L^2(G)$ be the Hilbert space of square integrable (with respect to μ) functions on G; for f and g in $L^2(G)$ we write

$$(f|g) = \int f(x)\overline{g(x)}d\mu(x).$$

The matrix elements of (unitary) irreducible representations form an orthogonal basis of $L^2(G)$; we have also

$$(\chi|\chi') = \begin{cases} 0, & \chi \neq \chi' \\ 1, & \chi = \chi' \end{cases}$$

for any two irreducible characters χ and χ'. Let H be a subgroup of finite index $d(H) = [G:H]$; obviously,

$$d(H)\mu(H) = 1. \tag{1}$$

Given a representation $B: H \to GL(m,\mathbb{C})$, we let $B(x) = 0$ for $x \in G \setminus H$ and define a representation

$$A: G \to GL(nm,\mathbb{C}), \quad n := d(H),$$

by the relation

$$A(x) = \begin{pmatrix} B(t_1xt_1^{-1}) & \dots & B(t_1xt_n^{-1}) \\ \dots\dots & \dots\dots & \dots\dots \\ B(t_nxt_1^{-1}) & \dots & B(t_nxt_n^{-1}) \end{pmatrix} ,$$

where $\{t_j | 1 \leq j \leq n\}$ is a set of representatives for the right cosets of H in G, so that $G = \bigcup_{j=1}^{n} Ht_j$ and $Ht_j \neq Ht_i$ for $i \neq j$. One writes $A = \mathrm{Ind}_H^G(B)$; we denote the character of A by χ^G, where χ denotes the character of B. Obviously,

$$\chi^G(x) = \sum_{j=1}^{n} \chi(t_jxt_j^{-1}). \tag{2}$$

<u>Lemma 1</u>. The character of a finite dimensional representation of G determines this representation (up to equivalence).

<u>Proof</u>. The characters of two different irreducible representations having different decompositions in the basis of matrix elements of irreducible representations can not coincide. Since every representation of a compact group can be decomposed in a direct sum of irreducible ones, we see that a character determines its representation.

<u>Lemma 2</u>. Let ψ be a character of H. Then

$$\mu(H)\psi^G(x) = \int_G \psi(yxy^{-1})d\mu(y), \quad x \in G. \tag{3}$$

<u>Proof</u>. We have

$$\int_G \psi(yxy^{-1})dy = \sum_{i=1}^{n} \int_H \psi(ut_ixt_i^{-1}u^{-1})du. \tag{4}$$

But $y \in H$ if and only if $uyu^{-1} \in H$ for $u \in H$, therefore $\psi(uyu^{-1}) = \psi(y)$ whenever $u \in H$. Thus (4) gives

$$\int_G \psi(yxy^{-1})dy = \sum_{i=1}^{n} \int_H \psi(t_i x t_i^{-1})du = \mu(H)\psi^G(x).$$

Let φ be a character of G. We denote by φ_H the restriction of φ to H (sometimes we write $(\varphi|H)$ for φ_H).

Proposition 1. Let ψ be a character of H and let φ be a character of G. Then

$$\mu(H) <\psi^G|\varphi> = <\psi|\varphi_H> . \tag{5}$$

Proof. Write

$$I = \int_{G\times G} \psi(yxy^{-1})\overline{\varphi(x)}d\mu(x)d\mu(y).$$

By lemma 2, $I = \mu(H) <\psi^G|\varphi>$. On the other hand,

$$I = \int_{G\times G} \psi(x)\overline{\varphi(y^{-1}xy)}d\mu(x)d\mu(y) = <\psi|\varphi_H>,$$

since $\varphi(y^{-1}xy) = \varphi(x)$ for $x \in G$, $y \in G$.

Remark 1. Equation (5), in view of (1), may be rewritten as follows: $<\psi^G|\varphi> = <\psi|\varphi_H>_H$, where $<\cdot|\cdot>_H$ denotes the scalar product in $L^2(H)$.

Lemma 3. If H is an abelian subgroup of G, then dimension of an irreducible representation of G doesn't exceed $[G:H]$.

Proof. Let χ be a simple character of G and let

$$\chi_H = \sum_{j=1}^{m} a_j\chi_j, \quad a_j \in \mathbb{N} ,$$

be the decomposition of χ_H into simple characters of H. Suppose that

$\chi_j \neq \chi_i$ for $j \neq i$ and that, say, $a_1 > 0$. By Proposition 1,

$$\langle \chi_1^G | \chi \rangle = \langle \chi_1 | \chi_H \rangle \quad .$$

In view of orthogonality relations, $\langle \chi_1 | \chi_H \rangle_H = a_1$, so that $\langle \chi_1^G | \chi \rangle = a_1$. Since χ is a simple character, we have

$$\chi_1^G = a_1 \chi + \psi \ , \qquad \langle \chi | \psi \rangle = 1. \tag{6}$$

On the other hand, χ_1^G is of degree $[G:H]$ because χ_1 is of degree one (H is abelian!), therefore (6) gives

$$\chi(1)a_1 \leq [G:H], \quad \chi(1) \leq [G:H].$$

Let A and B be two subgroups of finite index in G and let $G = \bigcup_{i=1}^{n} At_iB$ be a decomposition of G in double cosets modulo (A,B), so that $At_iB \cap At_jB = \emptyset$ for $i \neq j$. Suppose, moreover, that

$$C_i = t_i^{-1}At_i \cap B, \qquad 1 \leq i \leq n,$$

is a subgroup of finite index in G.

<u>Proposition 2</u>. Let ρ be a representation of A and let θ be a representation of B. Then

$$\mathrm{Ind}_A^G(\rho) \otimes \mathrm{Ind}_B^G(\theta) = \sum_{i=1}^{n} \oplus \ \mathrm{Ind}_{C_i}^G(\sigma_i), \tag{7}$$

where $\sigma_i = \rho_i \otimes \theta_i$, $\rho_i : x \to \rho(t_i x t_i^{-1})$, $\theta_i : x \to \theta(x)$ are representations of C_i, $1 \leq i \leq n$.

<u>Proof</u>. Let $\varphi = \mathrm{tr}\ \rho$, $\chi = \mathrm{tr}\ \theta$ and $\xi_i = \mathrm{tr}\ \sigma_i$ be the characters

the representations ρ, θ and σ_i, respectively. In view of lemma 1, it is enough to show that

$$\varphi^G \psi^G = \sum_{i=1}^{n} \xi_i^G . \tag{8}$$

We have $\xi_i(x) = \varphi(t_i x t_i^{-1})\psi(x)$ for $x \in C_i$. By lemma 2,

$$\mu(C_i)\xi_i^G(x) = \int_G \xi_i(uxu^{-1})d\mu(u) = \int_G \xi_i(vuxu^{-1}v^{-1})d\mu(u) \tag{9}$$

with $v \in B$. Taking into account the obvious identities:

$$\psi(vxv^{-1}) = \psi(x), \quad \varphi(wxw^{-1}) = \varphi(x), \quad v \in A, \quad w \in B$$

one obtains from (9) an equation

$$\mu(C_i)\xi_i^G(x) = \int_G d\mu(u)\psi(uxu^{-1})\,\varphi((wt_i v)uxu^{-1}(wt_i v)^{-1}) \tag{9'}$$

for any v, w such that $(v,w) \in A \times B$. Let

$$f_A(x) = \begin{cases} 1, & x \in A \\ 0, & x \notin A \end{cases} \qquad \text{and} \qquad f_B(x) = \begin{cases} 1, & x \in B \\ 0, & x \notin B \end{cases}$$

be the characteristic functions of A and B, and consider an integral

$$J_i(x) := \int_{G\times G} f_A(y) f_B(z)\varphi((yt_i z)x(yt_i z)^{-1})d\mu(y)d\mu(z).$$

Integration over $A\times B$ in the both sides of (9') gives:

$$\mu(C_i)\mu(A)\mu(B)\xi_i^G(x) = \int_G \psi(uxu^{-1})J_i(uxu^{-1})d\mu(u). \tag{10}$$

Let $T_i = At_iB$ and let $g_x(u) = \varphi(uxu^{-1})$. One obtains:

$$J_i(x) = \int_{G\times G} f_A(y)f_B(z)g_x(yt_iz)d\mu(y)d\mu(z),$$

so that a translation $y \mapsto yt_iz$ gives

$$J_i(x) = \int_{T_i} g_x(u)d\mu(u) \int_B f_A(uz^{-1}t_i^{-1})d\mu(z).$$

We remark now that

$$\int_B f_A(uz^{-1}t_i^{-1})d\mu(z) = \mu(B \cap u^{-1}At_i)$$

and that $\mu(B \cap u^{-1}At_i) = \mu(C_i)$ for $u \in T_i$. Thus

$$J_i(x) = \mu(C_i) \int_{T_i} \varphi(uxu^{-1})d\mu(u). \tag{11}$$

Equations (10) and (11) give

$$\mu(A)\mu(B)\xi_i^G(x) = \int_G \psi(uxu^{-1})d\mu(u) \int_{T_i} \varphi(vuxu^{-1}v^{-1})d\mu(v), \tag{12}$$

since $\mu(C_i) \neq 0$. Equation (8) follows from (12) and (3) when one sums over i and takes into account that $T_i \cap T_j = \emptyset$ for $i \neq j$. This proves Proposition 2.

If B is a one dimensional representation of H, we say that representation $A = \mathrm{Ind}_H^G(B)$ is <u>monomial</u>. It follows from Proposition 2, if its conditions are satisfied, that tensor product of monomial representations can be decomposed in a direct sum of monomial representations. The following statement is a particular case of this observation.

<u>Corollary 1</u>. Let H_i be a subgroup of G and let $H^{(j)} = \bigcap_{i=1}^{j} H_i$, $1 \leq j \leq r$. Suppose that $H^{(r)}$ is of finite index in G and that if

$1 \le j \le r-1$, then $G = H_{j+1}H^{(j)}$. Consider r monomial representations $\rho_i = \mathrm{Ind}_{H_i}^G(\chi_i)$, $1 \le i \le r$, induced by one dimensional representations $\chi_i\colon H_i \to GL(1,\mathbb{C})$ and let $\chi^{(r)} = \prod_{i=1}^{r}(\chi_i|H^{(r)})$. Then

$$\rho_1 \otimes \dots \otimes \rho_r = \mathrm{Ind}_{H^{(r)}}^G(\chi^{(r)}). \tag{13}$$

Proof. Let $\chi^{(j)} = \prod_{i=1}^{j}(\chi_i|H^{(j)})$, $1 \le j \le r$. We prove that

$$\rho_1 \otimes \dots \otimes \rho_j = \mathrm{Ind}_{H^{(j)}}^G(\chi^{(j)}) \tag{14}$$

for $1 \le j \le r$. For $j = 1$ equation (14) is obvious. Suppose (14) holds for some j in the interval $1 \le j \le r-1$. Since

$$\mathrm{Ind}_{H^{(j)}}^G(\chi^{(j)}) \otimes \mathrm{Ind}_{H_{j+1}}^G(\chi_{j+1}) = \mathrm{Ind}_{H^{(j+1)}}^G(\chi^{(j+1)})$$

by Proposition 2 (in view of the condition $H^{(j)}J_{j+1} = G$), equation (14) implies that

$$\rho_1 \otimes \dots \otimes \rho_{j+1} = \mathrm{Ind}_{H^{(j+1)}}^G(\chi^{(j+1)}).$$

This proves (14) for any j, in particular, we obtain (13).

If G is a finite group, the following relation holds:

$$\sum_{\chi\in\hat{G}} \chi(g_1)\overline{\chi(g_2)} = \begin{cases} 0 & \text{when } g_1 \notin \{g_2\} \\ \dfrac{|G|}{|\{g_1\}|} & \text{when } g_1 \in \{g_2\} \end{cases}, \tag{15}$$

where $\{g\} = \{hgh^{-1} | h \in G\}$ denotes the conjugacy class of g in G.

Theorem 1. Every character of a finite group is a linear combination of monomial characters.

Proof. See, e.g., [83], §10.

Lemma 4. Let $A \in GL(n,\mathbb{C})$. If A is semisimple, then

$$\det(I-At)^{-1} = \sum_{m=0}^{\infty} t^m \operatorname{tr}(S^m A), \tag{16}$$

and

$$\det(I-At) = \sum_{m=0}^{\infty} (-1)^m \operatorname{tr}(\Lambda^m A) t^m =$$

$$\sum_{m=0}^{n} (-1)^m \operatorname{tr}(\Lambda^m A) t^m, \tag{17}$$

where $S^m A$ and $\Lambda^m A$ denote the m-th symmetric and exterior powers of A respectively. Identities (16) and (17) are understood to hold formally in $\mathbb{C}[[t]]$.

Proof. Let V be an n-dimensional complex vector space and let $\{\ell_1,\ldots,\ell_n\}$ be a $\mathbb{C}$-basis of V. Suppose that $A\ell_i = \alpha_i \ell_i$, $1 \le i \le n$. Then

$$\det(1-At)^{-1}(\ell_1 \wedge \ldots \wedge \ell_n) = (1-At)^{-1}\ell_1 \wedge \ldots \wedge (1-At)^{-1}\ell_n \tag{18}$$

and

$$(1-At)^{-1}\ell_i = \sum_{m=0}^{\infty} \alpha_i^m t^m \ell_i, \quad 1 \le i \le n. \tag{19}$$

Identities (18) and (19) give:

$$\det(1-At)^{-1} = \sum_{m=0}^{\infty} t^m \sum_{m_1+\ldots+m_n=m} \alpha_1^{m_1} \ldots \alpha_n^{m_n},$$

and (16) follows. Identity (17) follows from the equations:

$$\det(1+At)\ell_1 \wedge \ldots \wedge \ell_n = (1+At)\ell_1 \wedge \ldots \wedge (1+At)\ell_n$$

and

$$(1+At)\ell_j = (1+\alpha_j t)\ell_j \quad , \qquad 1 \leq j \leq n.$$

§3. Weil's groups and non-abelian L-functions.

Let K be a finite Galois extension of k. The relative Weil group $W(K|k)$ is defined as the extension of $G(K|k)$ by C_K determined by the fundamental class of class field theory. If $K \supseteq K'$ and $K'|k$ is a Galois extension, then

$$W(K'|k) \cong W(K|k)/W(K|K')^c \quad , \tag{1}$$

where $W(K|K')$ is identified with a subgroup of $W(K|k)$. Thus one can define the absolute Weil group $W(k)$ as the projective limit

$$W(k) = \varprojlim W(K|k) \tag{2}$$

over finite Galois extensions $K|k$. In view of (2), any continuous finite dimensional representation

$$\rho\colon W(k) \to GL(m,\mathbb{C}) \tag{3}$$

of $W(k)$ factors through $W(K|k)$ for a finite Galois extension $K|k$. Parallel to the decomposition $C_K = \mathbb{R}_+ \times C_K^1$, we have a decomposition

$$W(K|k) = \mathbb{R}_+ \times W_1(K|k), \tag{4}$$

where $W_1(K|k)$ is defined as an extension of $G(K|k)$ by C_K^1. In particular, $W_1(K|k)$ is compact. A representation (3) is said to be <u>normalised</u> if ρ factors through $W_1(K|k)$ for a finite Galois extension $K|k$. Let $R(k)$ denote the set of all the normalised (continuous finite dimensional) representations of $W(k)$. Let $R(K|k)$ be the subset of $R(k)$ consisting of all the representations that factor through $W_1(K|k)$.

<u>Proposition 1</u>. Let $\rho \in R(K|k)$ and suppose that ρ is irreducible, then

$$\dim \rho \leq [K:k]. \tag{5}$$

Proof. Since $W_1(K|k)$ is compact and $[W_1(K|k):C_K^1] = [K:k]$ relation (5) follows from lemma 2.3.

For each p in $S_o(k)$ let $\hat{k}_p$ be a fixed algebraic closure of k_p; let $k_p \leq K_{\mathfrak{p}} \leq \hat{k}_p$ with $\mathfrak{p} \in S_o(K)$. Consider the maximal abelian extension $K_{\mathfrak{p}}^a|K_{\mathfrak{p}}$ in $\hat{k}_p$ and denote by $v_{\mathfrak{p}}^a$ and $\mathfrak{p}^a$ the ring of integers and the prime divisor of $K_{\mathfrak{p}}^a$, respectively. The groups $K_{\mathfrak{p}}^*$ and $G(K_{\mathfrak{p}}|k_p)$ can be regarded as subgroups of C_k and $G(K|k)$, respectively; identifying $K_{\mathfrak{p}}^*$ with $G(K_{\mathfrak{p}}^a|K_{\mathfrak{p}})$ we may regard $G(K_{\mathfrak{p}}^a|k_p)$ as a subgroup of $W(K|k)$. This subgroup is denoted by $W(K_{\mathfrak{p}}|k_p)$; thus $W(K_{\mathfrak{p}}|k_p) \cong G(K_{\mathfrak{p}}^a|k_p)$. One defines two subsets of $W(K_{\mathfrak{p}}|k_p)$:

$$\iota(p) = \{\sigma | \sigma \in W(K_{\mathfrak{p}}|k_p),\quad \alpha^\sigma \equiv \alpha(\mathfrak{p}^a) \quad \text{for} \quad \alpha \in v_{\mathfrak{p}}^a\}$$

and

$$\sigma_p = \{\sigma | \sigma \in W(K_{\mathfrak{p}}|k_p),\quad \alpha^\sigma \equiv \alpha^{|p|}(\mathfrak{p}^a) \quad \text{for} \quad \alpha \in v_{\mathfrak{p}}^a\}.$$

Let us recall that

$$W(K_{\mathfrak{p}}|k_p) = \bigcup_{n=-\infty}^{\infty} \sigma_p^n \quad , \quad \sigma_p^o = \iota(p),$$

so that

$$W(K_{\mathfrak{p}}|k_p)/\iota(p) \cong \mathbb{Z} \ .$$

Let $\rho \in R(K|k)$ and let V denote the representation space of ρ; consider a subspace

$$V_p = \{x \mid x \in V,\ \rho(\tau)x = x \quad \text{for} \quad \tau \in \iota(p)\}, \qquad p \in S_o(k),$$

of V. We say that ρ is <u>unramified at</u> p if $V_p = V$.

<u>Proposition 2</u>. Let $\rho \in R(K|k)$ and let $p \in S_o(k)$. If p is unramified in $K|k$ and if $U_{\mathfrak{p}} \leq \operatorname{Ker} \rho$ whenever $\mathfrak{p}|p$, $\mathfrak{p} \in S_o(K)$, then ρ is unramified at p. Here $U_{\mathfrak{p}}$ denotes the subgroup of units in $K_{\mathfrak{p}}^*$ regarded as a subgroup of C_K.

<u>Proof</u>. Let $\tau \in \iota(p)$. Then $\alpha^\tau \equiv \alpha(\mathfrak{p})$ for $\alpha \in v_{\mathfrak{p}}^a \cap K_{\mathfrak{p}}$; since $K_{\mathfrak{p}}|k_p$ is unramified, we have $\alpha^\tau = \alpha$ for $\alpha \in K_{\mathfrak{p}}$. Thus

$$\iota(p) \leq G(K_{\mathfrak{p}}^a|K_{\mathfrak{p}}). \tag{6}$$

By virtue of local class field theory, it follows from (6) that $\iota(p) = U_{\mathfrak{p}}$; since $U_{\mathfrak{p}} \leq \operatorname{Ker} \rho$, we conclude that ρ is unramified at p.

<u>Notation 1</u>. Let $\rho \in R(k)$. We denote by $S_o(\rho)$ the set of those primes in $S_o(k)$ at which ρ is ramified.

<u>Proposition 3</u>. The set $S_o(\rho)$ is finite for any ρ in $R(k)$.

<u>Proof</u>. Suppose that $\rho \in R(K|k)$. The restriction of ρ to C_K is a continuous homomorphism of C_K in $GL(n,\mathbb{C})$ for some n in $\mathbb{N}$; therefore there is a finite subset $S_3(\rho)$ of primes in $S_o(K)$ such that

$$U_{\mathfrak{p}} \leq \operatorname{Ker} \rho \quad \text{for} \quad \mathfrak{p} \in S_o(K) \setminus S_3(\rho). \tag{7}$$

By Proposition 2 and (7),

$$S_o(\rho) \leq \{p \mid \mathfrak{p} \in S_o(k),\ \ \mathfrak{p}|p \ \text{ for some } \ \mathfrak{p} \text{ in } S_3(\rho)\}$$

$$\cup\ \{p \mid p \in S_o(k),\ \ p \ \text{ is ramified in } \ K|k\}.$$

Therefore the set $S_o(\rho)$ is finite.

Notation 2. Let $\rho \in R(K|k)$. We denote by $\rho(\sigma_p)$ the restriction of $\rho(\tau)$ to V_p for $\tau \in \sigma_p$, $p \in S_o(k)$.

The operator

$$\rho(\sigma_p): V_p \to V_p$$

is well defined, since the restriction of $\rho(\tau)$ to V_p doesn't depend on the choice of τ in σ_p. Let $\rho \in R(k)$; choose $K|k$ so that $\rho \in R(K|k)$ and let

$$\ell_p(\rho,t) = \det(1-t\rho(\sigma_p)), \quad p \in S_o(k). \tag{8}$$

Obviously, definition of $\ell_p(\rho,t)$ doesn't depend on the choice of the extension $K|k$. It follows from (1) that $W(K|k)^a \cong C_k$, therefore any character in $gr(k)$ may be regarded as an element of $R(k)$; with obvious identifications, one has a relation

$$gr(k) \subseteq \bigcap_{K|k} R(K|k), \tag{9}$$

where $K|k$ ranges over finite Galois extensions of k.

Proposition 4. The polynomials $\ell_p(t,\rho)$ have the following properties:

1) $\ell_p(\rho_1 \oplus \rho_2,t) = \ell_p(\rho_1,t)\ell_p(\rho_2,t)$ for ρ_1,ρ_2 in $R(k)$; (10)

2) if $\rho \in gr(k)$ and $\chi = \mathrm{tr}\,\rho$, then

$$\ell_p(\rho,t) = 1-t\chi(p); \tag{11}$$

3) if $\rho = \mathrm{Ind}_{W(k')}^{W(k)}\rho'$ with $\rho' \in R(k')$, then

$$\ell_p(\rho,t) = \prod_{\mathfrak{p}|p} \ell\,(\rho',t^{f(\mathfrak{p})}), \tag{12}$$

where $\mathfrak{p}$ range over primes in $S_o(k')$ lying above p, and $N_{k'/k}\mathfrak{p} = p^{f(\mathfrak{p})}$; here $k'|k$ is a finite field extension.

Proof. Assertion 2) is a reformulation of the Artin's reciprocity law: equation (11) follows from the definitions when one recalls that the inertia subgroup and the Frobenius class in $G(k_p^a|k_p)$ may be identified with U_p and πU_p in C_k, where $\pi \in k_p^*$ and $w_p(\pi) = 1$. Let us remark that given a divisor $\mathfrak{P}$ in $S_o(K)$ such that $\mathfrak{P}|p$ and a representation

$$\rho_p : W(K_{\mathfrak{P}}|k_p) \to GL(\ell,\mathbb{C}), \quad \ell \in \mathbb{N},$$

one can define a local factor

$$\ell_p(\rho_p,t) = \det(1-\rho_p(\sigma_p)t),$$

where $\rho_p(\sigma_p)$ is defined as the restriction of $\rho_p(\tau)$ for $\tau \in \sigma_p$ to V_p, the subspace of $\iota(p)$-invariant vectors in the representation space of ρ_p. If, in particular,

$$\rho_p = \rho|_{W(K_{\mathfrak{P}}|k_p)} \quad \text{for} \quad \rho \in R(K|k),$$

then $\ell_p(\rho_p,t) = \ell_p(\rho,t)$ independently of the choice of $\mathfrak{P}$ above p. Given two representations $\rho_p^{(1)}$ and $\rho_p^{(2)}$ of $W(K_{\mathfrak{P}}|k_p)$, we observe that

$$V_p^{(1)} \oplus V_p^{(2)} = V_p,$$

where V_p, $V_p^{(1)}$ and $V_p^{(2)}$ denote the subspaces of $\iota(p)$-invariant vectors in the representation spaces of $\rho_p, \rho_p^{(1)}$ and $\rho_p^{(2)}$, respectively; $\rho_p := \rho_p^{(1)} \oplus \rho_p^{(2)}$. Therefore

$$\ell_p(\rho_p^{(1)} \oplus \rho_p^{(2)}, t) = \ell_p(\rho_p^{(1)}, t)\ell_p(\rho_p^{(2)}, t). \tag{13}$$

Identity (10) is a special case of (13) with $\rho_p^{(i)} = \rho_i|_{W(K_{\mathfrak{P}}|k_p)}$, $i = 1,2$. To prove (12) we need the following lemma.

<u>Lemma 1</u>. Let $\rho' : W(K_{\mathfrak{P}}|k'_{\mathfrak{p}}) \to GL(\ell,\mathbb{C})$, $\ell \in \mathbb{N}$, be a representation of $W(K_{\mathfrak{P}}|k'_{\mathfrak{p}})$ and let

$$\rho_{\mathfrak{p}} = \mathrm{Ind}_{W(K_{\mathfrak{P}}|k'_{\mathfrak{p}})}^{W(K_{\mathfrak{P}}|k_p)} \rho' .$$

Then

$$\ell_p(\rho_{\mathfrak{p}}, t) = \ell_{\mathfrak{p}}(\rho'_{\mathfrak{p}}, t^f), \tag{14}$$

where $N_{k'|k}\mathfrak{p} = p^f$.

<u>Proof</u>. Let $\iota(\mathfrak{p}) = \iota(\mathfrak{P}) \cap W(K_{\mathfrak{P}}|k'_{\mathfrak{p}})$ be the inertia subgroup of $W(K_{\mathfrak{P}}|k'_{\mathfrak{p}})$ and let

$$W(K_{\mathfrak{P}}|k_p) = \bigcup_{e\in\varepsilon} \bigcup_{n=1}^{f} W(K_{\mathfrak{P}}|k'_{\mathfrak{p}})e\tau^n, \tag{15}$$

where $\tau \in \sigma_p$ and ε is a (finite) set of representatives in $\iota(p)$ modulo $\iota(\mathfrak{p})$. Let $\chi = \mathrm{tr}\ \rho_{\mathfrak{p}}$ and $\chi' = \mathrm{tr}\ \rho'_{\mathfrak{p}}$. By (15) and (2.2),

$$\chi(\sigma) = \sum_{e\in\varepsilon} \sum_{n=1}^{f} \chi'(e\tau^n\sigma\tau^{-n}e^{-1}), \qquad \sigma \in W(K_{\mathfrak{P}}|k_p). \tag{16}$$

Let

$$a_m = \int_{\iota(p)} \chi(\tau^m u)\, d\mu(u).$$

Since $u \to \tau^n u \tau^{-n}$ is an automorphism of $\iota(p)$, one obtains from (16) an equation:

$$a_m = f \sum_{e \in \varepsilon} \int_{\iota(p)} \chi'(e\tau^m u e^{-1})\, d\mu(u).$$

Changing the variable of integration (at first by $u \to ue^{-1}$, then by $u \to \tau^m u \tau^{-m}$ and finally by $u \to eu$) we can rewrite this equation as follows:

$$a_m = f|\varepsilon| \int_{\iota(p)} \chi'(u\tau^m)\, d\mu(u), \tag{17}$$

where μ is the Haar measure on $\iota(p)$ normalised by the condition $\mu(\iota(p)) = 1$. By definition, $\chi'(v) = 0$ for $v \notin W(K_{\mathfrak{p}}|k'_{\mathfrak{p}})$. Since $u\tau^m \in W(K_{\mathfrak{p}}|k'_{\mathfrak{p}})$ only when $f|m$ and $u \in \iota(\mathfrak{p})$, it follows from (17) that

$$a_m = \begin{cases} 0 & \text{if } f\nmid m \\ f \int_{\iota(\mathfrak{p})} \chi'(u\tau^m)\, d\mu'(u) & \text{if } f|m \end{cases}, \tag{18}$$

where μ' denotes the Haar measure on $\iota(\mathfrak{p})$ for which $\mu'(\iota(\mathfrak{p})) = 1$ (so that $d\mu' = |\varepsilon|\, d\mu$). On the other hand,

$$\log \ell_p(\rho_p, t)^{-1} = \sum_{m=1}^{\infty} \frac{t^m}{m} \operatorname{tr}(\rho(\tau^m)|V_p), \quad \tau \in \sigma_p. \tag{19}$$

Since

$$\operatorname{tr}(\rho(\gamma)|V_p) = \operatorname{tr}(\rho(\gamma) \cdot \int_{\iota(p)} \rho(u)\, d\mu(u)) = \operatorname{tr} \int_{\iota(p)} \rho(\gamma u)\, d\mu(u)$$

for any γ in $W(K_{\mathfrak{P}}|k_p)$ and any (finite dimensional continuous) representation ρ of $W(K_{\mathfrak{P}}|k_p)$, identity (19) may be rewritten as follows:

$$\log \ell_p(\rho_p, t)^{-1} = \sum_{m=1}^{\infty} a_m \frac{t^m}{m} . \tag{20}$$

Equations (18) and (20) give:

$$\log \ell_p(\rho_p, t)^{-1} = \sum_{m=1}^{\infty} b_m \frac{t^{fm}}{m} , \tag{21}$$

where

$$b_m = \int_{\iota(\mathfrak{p})} \chi'(\tau^m u)\, d\mu'(u), \qquad \tau \in \sigma .$$

In view of (20) applied to ρ', identity (14) follows from (21).

Let now

$$\rho'_{\mathfrak{p}} = \rho' | W(K_{\mathfrak{P}}|k'_{\mathfrak{p}})$$

and let

$$\rho_{\mathfrak{p}} = \mathrm{Ind}_{W(K_{\mathfrak{P}}|k'_{\mathfrak{p}})}^{W(K_{\mathfrak{P}}|k_p)} \rho'_{\mathfrak{p}} .$$

An easy calculation shows that

$$\rho | W(K_{\mathfrak{P}}|k_p) = \sum_{\mathfrak{p}|p} \oplus \rho_{\mathfrak{p}} , \qquad \mathfrak{p} \in S_o(k'). \tag{22}$$

Identity (12) is a consequence of (14), (22) and (13). This completes the proof of Proposition 4.

Let $\rho \in R(k)$. We say that ρ is of <u>Galois type</u> if $\mathrm{Im}\,\rho$ is finite; let $R_o(k)$ be the set of representations of Galois type in $R(k)$ and

let

$$R_o(K|k) = \{\rho \in R(K|k) \mid C_K \leq \mathrm{Ker}\ \rho\} .$$

Lemma 2. If ρ is of Galois type, then $\rho \in R_o(K|k)$ for a finite Galois extension $K|k$.

Proof. Suppose that ρ is of Galois type and that $\rho \in R(K'|k)$ for a finite Galois extension $K'|k$. Then there is a finite abelian extension $K|K'$ such that $N_{K/K'}(C_K) \leq \mathrm{Ker}\ \rho$; therefore ρ factors through $G(K|k)$. Thus $\rho \in R_o(K|k)$.

Corollary 1. Every closed subgroup of finite index in $W(k)$ is of the shape $W(k')$ for a finite Galois extension $k'|k$.

Proof. Since $G(k'|k) \cong W(k)/W(k')$, the argument used in the proof of Lemma 2 suffices to establish this statement.

A representation ρ in $R(k)$ is said to be primitive, if ρ is not induced by any other representation in $R(k)$.

Proposition 5. Let $\rho \in R(k)$ and suppose that ρ is primitive and irreducible. Then $\rho = \rho_1 \otimes \rho_2$ with $\rho_1 \in gr(k)$ and $\rho_2 \in R_o(k)$.

Proof. See [87], p. 10, and [91].

We remark that $R_o(K|k) \leq R_o(k) \cap R(K|k)$. Let $\rho \in R(k)$ and let $\chi = \mathrm{tr}\ \rho$. One defines the Artin-Weil L-function associated to ρ by an Euler product

$$L(s,\chi) = \prod_{p \in S_o(k)} \ell_p(\rho, |p|^{-s})^{-1} \tag{23}$$

The product (23) converges absolutely for $\mathrm{Re}\, s > 1$, $s \in \mathbb{C}$. It follows from (10) that

$$L(s,\chi_1+\chi_2) = L(s,\chi_1)L(s,\chi_2). \tag{24}$$

If ρ is an one-dimensional representation, then the definitions (23) and (1.8) agree, in view of (11) and (1.10). Finally, if

$$\rho = \mathrm{Ind}_{W(k')}^{W(k)}\rho' , \quad \rho' \in R(k'), \quad \chi' = \mathrm{tr}\,\rho' ,$$

then, by (12),

$$L(s,\chi) = L(s,\chi'). \tag{25}$$

<u>Theorem 1</u>. Let $\rho \in R(k)$ and $\chi = \mathrm{tr}\,\rho$. We have

$$L(s,\chi) = \prod_{j=1}^{m} L(s,\chi_j)^{e_j} , \tag{26}$$

where $e_j \in \{-1,1\}$, $\chi_j \in gr(k_j)$, $k_j|k$ is a finite field extension, $1 \leq j \leq m$.

<u>Proof</u>. We prove this statement by induction on the degree of χ. If ρ is one-dimensional, it is obvious (in view of (11)). If χ isn't simple, we use (24) to reduce the problem to the characters of lower degree; if ρ is induced from a proper subgroup, corollary 1 allows to use (25) and to reduce the problem again to the one involving representation ρ' with $\dim \rho' < \dim \rho$. Thus after passing, if necessary, to a finite extension of k we may assume that ρ is primitive and irreducible. Then, in view of Proposition 5,

$$\rho = \rho_1 \otimes \rho_2, \quad \rho_1 \in gr(k), \quad \rho_2 \in R_o(K|k) \tag{27}$$

for a finite Galois extension $K|k$ (cf. Lemma 2). By Theorem 2.1,

$$\rho_2 = \sum_{j=1}^{m'} \oplus \operatorname{Ind}_{G(K|k_j)}^{G(K|k)} \psi_j - \sum_{j=m'+1}^{m} \oplus \operatorname{Ind}_{G(K|k_j)}^{G(K|k)} \psi_j \tag{28}$$

for some finite extensions k_j and grossencharacters ψ_j in $gr(k_j)$, $1 \leq j \leq m$. It follows from (11) that $\rho_1(\sigma_p) = \chi_1(p)$ with $\chi_1 = \operatorname{tr} \rho_1$, so that

$$\ell_p(\rho,t) = \det(1-\rho_2(\sigma_p)(\chi_1(p)t)). \tag{29}$$

By (12), (28) and (29), we have

$$\ell_p(\rho,t) = \prod_{j=1}^{m} \prod_{\substack{\mathfrak{p}|p \\ \mathfrak{p}\in S_o(k_j)}} (1-\psi_j(\mathfrak{p})(\chi_1(p)t)^{f(\mathfrak{p})})^{e_j}, \tag{30}$$

where

$$e_j = \begin{cases} 1 & \text{when } j \leq r_1 \\ -1 & \text{when } j > r_1 \end{cases}.$$

Equation (26) with $\chi_j = (\chi_1 \cdot N_{k_j/k})\psi_j$, $1 \leq j \leq m$, follows from (30).

<u>Corollary 2</u>. The function

$$s \mapsto L(s,\chi) \tag{31}$$

defined by (23) for $\operatorname{Re} s > 1$ can be meromorphically continued to the whole complex plane $\mathbb{C}$.

<u>Proof</u>. Since $L(s,\chi_j)$, $1 \leq j \leq m$, is meromorphic in $\mathbb{C}$, this statement follows from (26).

<u>Conjecture (Artin-Weil)</u>. Let $\rho \in R(k)$, $\chi = \operatorname{tr} \rho$. If ρ is irreducible

and $\rho \neq 1$, then the function (31) is holomorphic in $\mathbb{C}$.

<u>Generalised Riemann Hypothesis</u>. Let $\rho \in gr(k)$ and $\chi = \mathrm{tr}\,\rho$. Then

$$L(s,\chi) \neq 0 \quad \text{for} \quad \mathrm{Re}\,s > \frac{1}{2}\,. \tag{R32}$$

Suppose that $L(s,\chi)$ satisfies (26) and let

$$L_\infty(s,\chi) = \prod_{j=1}^{m} L_\infty(s,\chi_j)^{e_j}\,. \tag{33}$$

Let further

$$W(\chi) = \prod_{j=1}^{m} W(\chi_j)^{e_j}\,, \qquad a(\chi) = \prod_{j=1}^{m} a(\chi_j)^{e_j}\,. \tag{34}$$

It follows from (33), (34), (1.12) and (1.14) that the function

$$s \mapsto \Lambda(s,\chi) := L(s,\chi)L_\infty(s,\chi) \tag{35}$$

satisfies a functional equation

$$\Lambda(s,\chi) = W(\chi)a(\chi)^{\frac{1}{2}-s}\,\Lambda(1-s,\bar{\chi})\,; \quad |W(\chi)| = 1\,. \tag{36}$$

<u>Definition 1</u>. Let $\rho \in R(k)$ and $\chi = \mathrm{tr}\,\rho$. We say that ρ is of <u>AW type</u>, if the function (31) is holomorphic in $\mathbb{C}\setminus\{1\}$.

<u>Corollary 3</u>. The Artin-Weil conjecture is equivalent to the statement that each ρ in $R(k)$ is of AW type.

<u>Proof</u>. The conjecture clearly implies this statement. Conversely, suppose that $\rho \in R(k)$ and $\rho \neq 1$, ρ is irreducible. Let $\chi = \mathrm{tr}\,\rho$ and let

$$L(s,\chi) = \prod_{j=1}^{\ell} L(s,\chi_j)^{e_j} \prod_{j=\ell+1}^{m} \zeta_{k_j}(s)^{e_j}, \tag{37}$$

where $\chi_j \in gr(k_j)$, $\chi_j \neq 1$, $k_j|k$ is a finite field extension, $1 \leq j \leq m$. Let $\rho_j = \mathrm{Ind}_{W(k_j)}^{W(k)} \chi_j$, $1 \leq j \leq m$, where we let $\chi_j = 1$ for $j > r'$. It follows from (37) and (12) that

$$\ell_p(\rho,t) = \ell_p(\rho',t), \quad \rho' := \sum_{j=1}^{m} \oplus\, e_j\rho_j, \quad p \in S_o(k).$$

Therefore (8) gives

$$\chi(\sigma_p) = \chi'(\sigma_p), \quad p \in S_o(k), \quad \chi' := \mathrm{tr}\ \rho'. \tag{38}$$

By the corollary A1.1, it follows from (38) that

$$\rho = \sum_{j=1}^{m} \oplus\, e_j\rho_j.$$

On the other hand, one deduces from (2.5) that ρ_j does not contain the identical representation of $W(k)$ if $j \leq \ell$ and contains it exactly once if $j > \ell$. Therefore

$$\sum_{j=\ell+1}^{m} e_j = 0. \tag{39}$$

Suppose that ρ is of AW type, then the function (31) is holomorphic $\mathbb{C}\setminus\{1\}$ and one concludes from (37) and (39) that it is holomorphic in $\mathbb{C}$.

§4. On character sums extended over integral ideals.

Let $\rho \in R(k)$ and let $\chi = \operatorname{tr} \rho$. We write

$$L(s,\chi) = \sum_{\mathfrak{n} \in I_o(k)} a(\mathfrak{n},\chi)\, |\mathfrak{n}|^{-s}, \qquad \operatorname{Re} s > 1, \tag{1}$$

and let

$$A(x,\chi) = \sum_{\substack{|\mathfrak{n}|<x \\ \mathfrak{n} \in I_o(k)}} a(\mathfrak{n},\chi), \qquad x > 0. \tag{2}$$

In this paragraph we estimate $A(x,\chi)$ asymptotically as $x \to \infty$ assuming that ρ is of AW type.

Notations 1. Given two (formal) Dirichlet series $f(s) = \sum_{m=1}^{\infty} a_m m^{-s}$ and $g(s) = \sum_{m=1}^{\infty} b_m m^{-s}$, we write $f(s) << g(s)$ if $|a_m| \leq b_m$ for each m.

Let $d(\chi)$ denote the degree of χ and let $n := [k:\mathbb{Q}]$.

Lemma 1. We have

$$L(s,\chi) << \zeta(s)^{nd(\chi)} . \tag{3}$$

Proof. Since ρ is equivalent to an unitary representation, it follows that

$$\ell_p(\rho,t) = \prod_{j=1}^{d(\chi)} (1-\varepsilon_j(p)t), \quad |\varepsilon_j(p)| \in \{0,1\}, \quad p \in S_o(k).$$

Therefore

$$\ell_p(\rho,t)^{-1} = \sum_{m=0}^{\infty} a_m t^m \quad \text{with } |a_m| \leq b_m \ ,$$

where

$$\sum_{m=0}^{\infty} b_m t^m = (1-t)^{-d(\chi)}.$$

Thus

$$L(s,\chi) \ll \zeta_k(s)^{d(\chi)}. \tag{4}$$

Comparing the Euler factors of $\zeta_k(s)$ and $\zeta(s)$ one obtains a relation:

$$\zeta_k(s) \ll \zeta(s)^n. \tag{5}$$

Relation (3) follows from (4) and (5).

Corollary 1. Let $\eta > 0$. Then

$$|L(1+\eta,\chi)| \leq (1+\eta^{-1})^{nd(\chi)} \tag{6}$$

Proof. It follows from (3) and the obvious inequality

$$\zeta(1+\eta) \leq 1+\eta^{-1}. \tag{7}$$

We need to estimate the Γ-factors occuring in the functional equation (3.36).

Lemma 2. Suppose that $f(s)$ is regular in the strip

$$S(a,b) = \{s \mid a \leq \operatorname{Re} s \leq b\}$$

and that there are C,c for which

$$|f(s)| < C \exp(|\operatorname{Im} s|^c) \quad \text{whenever} \quad s \in S(a,b).$$

If $|f(b+it)| \leq 1$ and $|f(a+it)| \leq A|Q+a+it|^{\alpha}$ whenever $t \in \mathbb{R}$ with $\alpha \geq 0$, $Q + a > 0$, then

$$|f(u)| \leq (A|Q+u|^{\alpha})^{\frac{b-\mathrm{Re}\, u}{b-a}} \quad \text{for} \quad u \in S(a,b). \tag{8}$$

Proof. It is a special case of Theorem 2 in [80, p. 195].

Lemma 3. The following estimates hold:

$$\left|\frac{\Gamma(q+1-u)}{\Gamma(q+u)}\right| \leq |q+1+u|^{1-2\,\mathrm{Re}\, u} \quad \text{whenever} \quad |\mathrm{Re}\, u| \leq \frac{1}{2}, \tag{9}$$

$$\left|\frac{\Gamma(q+u)}{\Gamma(q+1-u)}\right| \leq 4 \quad \text{whenever} \quad -\frac{1}{4} \leq \mathrm{Re}\, u \leq \frac{1}{2}, \tag{10}$$

$$\left|\Gamma\left(\frac{q+1-u}{2}\right)\Gamma\left(\frac{q+u}{2}\right)^{-1}\right| \leq \left(\frac{1}{2}|1+u|\right)^{\frac{1}{2}-\mathrm{Re}\, u} \quad \text{for} \quad |\mathrm{Re}\, u| \leq \frac{1}{2}, \tag{11}$$

$$\left|\Gamma\left(\frac{q+u}{2}\right)^{-1}\Gamma\left(\frac{q+1-u}{2}\right)\right| \leq 4 \quad \text{for} \quad |\mathrm{Re}\, u| \leq \frac{1}{2}. \tag{12}$$

In (1) and (2) we assume that $q \geq 0$ and $2q \in \mathbb{Z}$, while in (3) and (4) it is assumed that $q \in \{0,1\}$.

Proof. Estimates (9), (11), (12) and estimate (10) with $q \neq \frac{1}{2}$ follow from (8) with $a = -\frac{1}{2}$, $b = \frac{1}{2}$ and the functional equation

$$\Gamma(s+1) = s\Gamma(s), \quad s \in \mathbb{C}.$$

Let $f(u) = \Gamma(\frac{1}{2}+u)\Gamma(\frac{3}{2}-u)^{-1}$. We have

$$|f(\tfrac{1}{2}+it)| = 1, \quad f(-\tfrac{1}{4}+it) = \Gamma(\tfrac{1}{4}+it)\Gamma(\tfrac{7}{4}-it), \quad t \in \mathbb{R}_+.$$

Therefore it follows from the properties of the Γ-function (cf., e.g., [1], p. 255-256) that

$$f(-\tfrac{1}{4}+it) \le \frac{\Gamma(\frac{1}{4})}{\Gamma(\frac{7}{4})} \le 4, \qquad t \in \mathbb{R}_+ \quad ,$$

and we deduce (10) from (8) with $a = -\frac{1}{4}$, $b = \frac{1}{2}$, $A = 4$, $\alpha = 0$.

Suppose that $L(s,\chi)$ satisfies (3.26) with

$$e_j = \begin{cases} 1 & \text{for } 1 \le j \le m' \\ -1 & \text{for } m' + 1 \le j \le m \end{cases} .$$

Let

$$\tilde{S}_1 = \bigcup_{j=1}^{m'} S_1(k_j), \quad \tilde{S}_2 = \bigcup_{j=1}^{m'} S_2(k_j), \quad \bar{n} = \sum_{j=m'+1}^{m} [k_j:\mathbb{Q}]. \qquad (13)$$

For $p \in S_1(k_j)$ we write a_p and t_p for $a_p(\chi_j)$ and $t_p(\chi_j)$ respectively, where χ_j is a grossencharacter in (3.26), $1 \le j \le m$.

Proposition 1. The following estimate holds:

$$\left|\frac{L_\infty(1-s,\bar{\chi})}{L_\infty(s,\chi)}\right| \le 12^{\bar{n}}\Big(\prod_{p\in \tilde{S}_1} |1+s+it_p| \prod_{p\in\tilde{S}_2} \Big|1+s+\frac{|a_p|+it_p}{2}\Big|^2\Big)^{\frac{1}{2}-\mathrm{Re}\, s} \qquad (14)$$

in the interval $-\frac{1}{4} \le \mathrm{Re}\, s \le \frac{1}{2}$.

Proof. Let $\chi \in gr(k)$. In view of (9) - (12), it follows from the definitions that

$$\left|\frac{L_\infty(1-s,\bar{\chi})}{L_\infty(s,\chi)}\right| \le \Big((2\pi)^{-n} \prod_{p\in S_1} |1+s+it_p(\chi)| \prod_{p\in S_2} \Big|1+s+\frac{|a_p|+it_p}{2}\Big|^2\Big)^{\frac{1}{2}-\mathrm{Re}\, s}$$

for $|\mathrm{Re} s| \le \frac{1}{2}$; (15)

$$\left| \frac{L_\infty(s,\chi)}{L_\infty(1-s,\bar{\chi})} \right| \leq 4^{n_2} 2^{r_2} \pi^{n(\frac{1}{2}-\mathrm{Re}\, s)} \quad \text{for} \quad -\frac{1}{4} \leq \mathrm{Re}\, s \leq \frac{1}{2} \quad . \tag{16}$$

Estimate (14) is easily seen to follow from (15), (16) and the definitions (3.33), (13).

Lemma 4. Let the real numbers a,b,Q,γ satisfy the conditions

$$-Q < a < b, \quad \gamma \leq 0.$$

Then there is a function $\varphi(s,Q)$ regular in the strip $S(a,b)$ and such that

$$|\varphi(a+it,Q)| = |Q+a+it|^\gamma, \quad |\varphi(b+it,Q)| = 1, \qquad t \in R,$$

and

$$|\varphi(s,Q)| \geq |Q+s|^{\gamma \frac{b-\mathrm{Re}\, s}{b-a}} \qquad \text{for} \quad s \in S(a,b).$$

Moreover, $\varphi(s,Q) = O(|\mathrm{Im}\, s|^c)$ for $s \in S(a,b)$ (with a real constant c).

Proof. It is a special case of Theorem 1 in [80, p. 192 - 193].

Proposition 2. Suppose that ρ is of AW type and let $\chi = \mathrm{tr}\, \rho$. If $0 < \eta \leq \frac{1}{4}$, then

$$|L(s,\chi)| \leq (1+\eta^{-1})^{nd(\chi)} 12^{\bar{n}} (3|\tfrac{1+s}{1-s}|)^{g(\chi)} b(s,\chi)^{1+\eta-\mathrm{Re}\, s} \tag{AW 17}$$

in the strip $-\eta \leq \mathrm{Re}\, s \leq 1+\eta$, where $g(\chi)$ denotes the multiplicity of the identical representation in ρ and

$$b(s,\chi):=\Big(a(\chi)\prod_{p\in\tilde{S}_1}|1+s+it_p|\prod_{p\in\tilde{S}_2}|1+s+\frac{|a_p|+it_p}{2}|^2\Big)^{\frac{1}{2}}. \tag{18}$$

Proof. Making use of Lemma 4 we construct two functions $\varphi_1(s,1)$ and $\varphi_2(s,Q)$ with the following properties:

$$\varphi_1(s,1) = O(|\mathrm{Im}\ s|^c), \quad \varphi_2(s,Q) = O(|\mathrm{Im}\ s|^c); \tag{19}$$

$$|\varphi_1(s,1)| \geq |1+s|^{-\frac{1}{2}(1+\eta-\mathrm{Re}\ s)}, |\varphi_2(s,Q)| \geq |Q+s|^{-(1+\eta-\mathrm{Re}\ s)} \tag{20}$$

for $s \in S(a,b)$ and $|\varphi_1(b+it)| = |\varphi_2(b+it)| = 1$,

$$\varphi_j(a+it,Q_j) = |Q_j+a+it|^{\gamma_j}, \qquad j = 1,2, \tag{21}$$

where $Q_1 = 1$, $Q_2 = Q$, $\gamma_1 = -\frac{1}{2}-\eta$, $\gamma_2 = 2\gamma_1$, $t \in \mathbb{R}$. We let

$$a = -\eta, \qquad b = 1+\eta,$$

and define a new function $F\colon S(a,b) \to \mathbb{C}$ by the equation

$$F(s) = L(s,\chi)\prod_{p\in\tilde{S}_1}\varphi_1(s+it_p,1)\prod_{p\in\tilde{S}_2}\varphi_2\Big(s+i\frac{t_p}{2},1+\frac{|a_p|}{2}\Big)\varphi_3(s), \tag{22}$$

where

$$\varphi_3(s) = a(\chi)^{\frac{1}{2}(\mathrm{Re}\ s-1-\eta)}\,12^{\bar{n}(\mathrm{Re}\ s-1-\eta)}(1+2\eta)^{-1}. \tag{23}$$

By the functional equation (3.36), we have

$$|L(s,\chi)| = a(\chi)^{\frac{1}{2}-\mathrm{Re}\ s}\Big|\frac{L_\infty(1-s,\bar{\chi})}{L_\infty(s,\chi)}\Big|\cdot|L(1-s,\bar{\chi})|. \tag{24}$$

We remark that

$$|F(b+it)| = |L(1+\eta+it,\chi)| \leq (1+\eta^{-1})^{nd(\chi)}, \quad t \in R, \tag{25}$$

in view of (6) and since $|\varphi_j(b+it)| = 1, \quad 1 \leq j \leq 3$. By (21) - (23),

$$|F(a+it)| = 12^{-\bar{n}}\, b(-\eta+it,\chi)^{-1-2\eta} |L(-\eta+it,\chi)| \tag{26}$$

with $b(s,\chi)$ defined by (18). By (24),

$$|L(-\eta+it,\chi)| = a(\chi)^{\frac{1}{2}+\eta} \left|\frac{L_\infty(1+\eta-it,\bar{\chi})}{L_\infty(-\eta+it,\chi)}\right| \cdot |L(1+\eta-it,\bar{\chi})|. \tag{27}$$

Relations (26), (27), (14), (6) and (18) give:

$$|F(a+it)| \leq (1+\eta^{-1})^{nd(\chi)}, \quad t \in \mathbb{R}. \tag{28}$$

Suppose now that ρ doesn't contain the identical representation, then $F(s)$ satisfies conditions of Lemma 2 with $\alpha = 0$ in view of (25), (28) and (19). Therefore it follows that

$$|F(s)| \leq (1+\eta^{-1})^{nd(\chi)} \quad \text{for} \quad -\eta \leq \text{Re } s \leq 1+\eta. \tag{AW 29}$$

Inequality (17) (with $g(\chi) = 0$) follows from (29) in view of definitions (22), (23) and inequalities (20). To complete the proof of proposition 2 we make use of the following inequality (see, [80], p. 200, Theorem 4):

$$|\zeta_k(s)| \leq 3\left|\frac{1+s}{1-s}\right| \left(\sqrt{|D|}\, \frac{|1+s|}{2\pi}\right)^{n/2)^{1+\eta-\text{Re } s}} \zeta(1+\eta)^n \tag{30}$$

in the strip $-\eta \leq \text{Re} s \leq 1+\eta$. We write

$$L(s,\chi) = \zeta_k(s)^{g(\chi)} L(s,\chi') \tag{31}$$

(with $g(\chi') = 0$) and apply (30) to the first factor in (31). Since (17) has already been proved for representations which do not contain the identical one, we may apply it to the second factor in (31). This completes the proof of proposition 2.

Remark 1. In notations of (3.37) - (3.39), we have

$$\sum_{j=\ell+1}^{m} e_j = g(\chi). \tag{32}$$

Lemma 5. Suppose that the Dirichlet series

$$f(s) = \sum_{n=1}^{\infty} a_n n^{-s}$$

converges absolutely for $\operatorname{Re} s > 1$, that

$$\sum_{n=1}^{\infty} |a_n| n^{-\sigma} = O((\sigma-1)^{-\alpha}) \quad \text{for} \quad \sigma > 1$$

with $\alpha > 0$ and that

$$f(s) << \sum_{n=1}^{\infty} \psi(n) n^{-s}$$

for some non-decreasing function $\psi\colon \mathbb{R}_+ \to \mathbb{R}_+$. Then

$$\sum_{n<x} a_n = \frac{1}{2\pi i} \int_{c-iT}^{c+iT} f(w) \frac{x^w}{w} dw + O\left(\frac{x^c}{T(c-1)^\alpha}\right) + O\left(\frac{\psi(2x) x \log x}{T}\right) \tag{33}$$

where $c > 1$ and x lies in an interval $N + 1/4 < x < N + 1/2$, $N \in \mathbb{N}$. The implied by the O-symbol constants are effective and numerical.

Proof. See, e.g., [89], p. 53 - 55, lemma 3.12.

<u>Theorem 1</u>. Suppose that ρ is of AW type and let $\chi = \mathrm{tr}\,\rho$,

$$B(\chi) := 12^{\bar{n}} 6^{g(\chi)} (a(\chi)+1) \prod_{p \in \tilde{S}_1} (1+|t_p|) \prod_{p \in \tilde{S}_2} \left(1+\frac{|a_p|+|t_p|}{2}\right)^2. \qquad (34)$$

The following estimate holds:

$$A(x,\chi) = xP(\chi, \log x) + O_\varepsilon(B(\chi) x^{1-\frac{2}{N+2}+\varepsilon} (\log x)^{nd}), \qquad (\text{AW } 35)$$

where $N := \sum_{j=1}^{m'} [k_j : \mathbb{Q}]$, $P(\chi,t)$ is a polynomial of degree $g(\chi)-1$ when $g(\chi) \geq 1$ and $P(\chi,t) = 0$ when $g(\chi) = 0$, $x \geq 2$; $\varepsilon > 0$. Here the implied by O_ε-symbol constant may depend on ε and $nd(\chi)$.

<u>Proof</u>. By (3),

$$L(s,\chi) \ll \sum_{n=1}^{\infty} n^\varepsilon n^{-s} C_1(\varepsilon, nd(\chi)), \quad \varepsilon > 0 \qquad (36)$$

with an effectively computable (in terms of $nd(\chi)$ and ε) $C_1 > 0$. Therefore (33) gives:

$$A(x,\chi) = \frac{1}{2\pi i} \int_{1+\eta-iT}^{1+\eta+iT} L(w,\chi) \frac{x^w}{w}\, dw + O\left(\frac{x^{1+\eta}}{T\eta^{d(\chi)n}}\right) + O_\varepsilon\left(\frac{x^{1+\varepsilon} \log x}{T}\right).$$

Since ρ is of AW type, one can move the contour of integration to the line $\mathrm{Re}\, s = 0$ passing the residue of multiplicity $g(\chi)$ at $s = 0$. This procedure leads to the estimate (35) as soon as one takes $\eta = (\log x)^{-1}$, $T = x^{2/N+2}$ and replaces (17) by the estimate

$$|L(s,\chi)| \leq (1+\eta^{-1})^{nd(\chi)} B(\chi) (1+|\mathrm{Im}\, s|)^{N/2}. \qquad (\text{AW } 37)$$

§5. On character sums extended over prime ideals.

Let us assume at first that $\chi \in gr(k)$. In this case one deduces from (4.17) the following estimate:

$$|L(s,\chi)| \leq 3^n (3|\frac{1+s}{1-s}|)^{g(\chi)} b(s,\chi), \qquad \frac{1}{2} \leq \text{Re } s \leq \frac{3}{2}. \tag{1}$$

Let

$$\varphi(t,\chi) := (a(\chi) \prod_{p\in S_1} (3+|t|+|t_p(\chi)|) \prod_{p\in S_2} (3+|t|+\frac{|a_p(\chi)|+|t_p(\chi)|}{2})^2)^{1/2}$$

Proposition 1. There is a positive constant c_1 such that $L(s,\chi) \neq 0$ in the region

$$\text{Re } s > 1 - (c_1 \log \varphi(\text{Im } s, \chi))^{-1} \tag{2}$$

with a possible exceptional zero, which must be real and simple, when $\chi^2 = 1$.

Proof. If $\chi^2 = 1$ this assertion is proved in [39] (see Lemma 2.3 on p. 277). Suppose that $\chi^2 \neq 1$. Let

$$L(\alpha+it_o,\chi) = 0. \tag{3}$$

Without loss of generality we may assume that $t_o \geq 0$ and

$$\alpha \geq \frac{3}{4} + \beta, \qquad 0 < \beta \leq \frac{1}{2}. \tag{4}$$

Let

$$s_o = 1+\beta+it_o, \qquad s_1 = 1+\beta+i(2t_o). \tag{5}$$

Since

$$|L(s,\chi)^{-1}| \leq \zeta_k(\text{Re } s) \quad \text{for} \quad \text{Re } s > 1, \quad \chi \in gr(k),$$

it follows from (1) that one can find a positive constant c_2 satisfying three inequalities:

$$\left| \frac{L(s,\chi)}{L(s_o,\chi)} \right| \leq c_2^n \, \beta^{-1} \varphi(t_o,\chi) \quad \text{for} \quad |s-s_o| \leq \frac{1}{2} \,, \tag{6}$$

$$\left| \frac{L(s,\chi^2)}{L(s_1,\chi^2)} \right| \leq c_2^n \beta^{-1} \varphi(t_o,\chi) \quad \text{for} \quad |s-s_1| \leq \frac{1}{2} \,, \tag{7}$$

and

$$\left| \frac{\zeta_k(s)(s-1)}{\zeta_k(1+\beta)\beta} \right| \leq c_2^n \beta^{-2} a(1) \quad \text{for} \quad |s-1-\beta| \leq \frac{1}{2} \,. \tag{8}$$

By a classical theorem (see, e.g., [78], p. 384, Satz 4.4 and Satz 4.5), relations (6) - (8) imply:

$$\text{Re} \frac{L'(s_o,\chi)}{L(s_o,\chi)} \geq -8 \log(c_2^n \beta^{-1} \varphi(t_o,\chi)) + \frac{1}{1+\beta-\alpha} \,, \tag{9}$$

$$\text{Re} \frac{L'(s_1,\chi^2)}{L(s_1,\chi^2)} \geq -8 \log(c_2^n \beta^{-1} \varphi(t_o,\chi)), \tag{10}$$

and

$$\text{Re} \frac{\zeta_k'(1+\beta)}{\zeta_k(1+\beta)} \geq -8 \log(c_2^n \beta^{-2} a(1)) - \frac{1}{\beta} \,. \tag{11}$$

Inequalities (9) - (11) combined with an elementary inequality

$$3 \, \text{Re} \frac{\zeta'_k}{\zeta_k}(1+\beta) + 4 \, \text{Re} \frac{L'}{L}(s_o,\chi) + \text{Re} \frac{L'}{L}(s_1,\chi^2) \leq 0$$

give:

$$\frac{1}{1+\beta-\alpha} \leq \frac{3}{4}\,\beta^{-1} + 22 \log \beta^{-1} + A \tag{12}$$

with

$$A = 16 \log(c_2^n \varphi(t_0,\chi)). \tag{13}$$

Let $\beta = c_3 A^{-1}$ and choose c_3 in such a way that $c_3 > 0$ and that (12) implies an inequality

$$\alpha \geq 1 - \beta/6, \quad \beta = c_3 A^{-1}. \tag{14}$$

Relations (13) and (14) give

$$\alpha \geq 1 - (c_1 \log \varphi(t_0,\chi))^{-1}, \qquad c_1 > 0,$$

for a properly chosen c_1. This proves Proposition 1.

Let, for brevity,

$$f(s) = \log[L(s,\chi)\,(\tfrac{s-1}{s+1})^{g(\chi)}\,(\tfrac{s-\alpha}{s+\alpha})^{g_1(\chi)}]\,, \tag{15}$$

where $g_1(\chi) = -1$ if there is an α satisfying two conditions:

$$\alpha > 1 - (c_1 \log \varphi(0,\chi))^{-1}, \quad L(\alpha,\chi) = 0 \tag{16}$$

and $g_1(\chi) = 0$ when no α satisfies (16).

<u>Lemma 1</u>. There is a positive constant c_4 such that $f(s)$ is regular in the region

$$\text{Re } s > 1 - c_4(\log \varphi(\text{Im } s,\chi))^{-1} \tag{17}$$

and satisfies the following estimates:

$$\frac{f'}{f}(s) = O(\log^2 \varphi(\operatorname{Im} s,\chi)), \tag{18}$$

$$f(s) = O(\log(3^n|D|\log \varphi(\operatorname{Im} s,\chi))). \tag{19}$$

Proof. Let

$$\psi(t) = (2c_1 \log \varphi(t,\chi))^{-1}, \quad t \in \mathbb{R}, \tag{20}$$

and let

$$s_1(t) = 1 + \psi(t) + it, \quad t \in \mathbb{R}. \tag{21}$$

Let $R(t)$ and $r(t)$ be chosen so that

$$2\psi(t) \geq R(t) > r(t) \geq \frac{4}{3}\psi(t), \; R(t) - r(t) \geq c_5 R(t), \tag{22}$$

and the distance of α from the circle $|s-s_1(t)| = R(t)$ is larger than $c_6\psi(t)$, when $g_1(\chi) = -1$, for two positive constants c_5 and c_6. By a classical lemma (see, e.g., [78], p. 383, Satz 4.2), it follows from proposition 1 and definitions (20) - (22) that

$$|f(s)-f(s_1(t_1))| \leq 2(M - \operatorname{Re} f(s_1(t_1))) \frac{r(t_1)}{R(t_1)-r(t_1)}, \tag{23}$$

and

$$\left|\frac{f'}{f}(s)\right| \leq \frac{2R(t_1)}{(R(t_1)-r(t_1))^2} (M - \operatorname{Re} f(s_1(t_1))), \tag{24}$$

where $|s-s_1(t_1)| \leq r(t_1)$ and $M := \sup(\operatorname{Re} f(s))$ on the circle $|s-s_1(t_1)| = R(t_1)$. One remarks that for $\chi \in gr(k)$ relation (4.17) may be rewritten as follows:

$$|L(s,\chi)(\frac{1-s}{1+s})^{g(\chi)}| \leq \zeta_k(1+\eta)3^{g(\chi)}\varphi(\mathrm{Im}\ s,\chi)^{1+\eta-\mathrm{Re}\ s}, \qquad (25)$$

where $-\eta \leq \mathrm{Re}\ s \leq 1+\eta$, $0 < \eta \leq \frac{1}{2}$. By (15),

$$\mathrm{Re}\ f(s) = \log|L(s,\chi)(\frac{s-1}{s+1})^{g(\chi)}| + g_1(\chi)\log|\frac{s-\alpha}{s+\alpha}| . \qquad (26)$$

If $|s-s_1(t_1)| = R(t_1)$, it follows from (25), (26) and the definition of $R(t)$ that

$$\mathrm{Re}\ f(s) \leq \log\zeta_k(1+\eta) + \log\psi(t_1)^{-1} + c_7, \qquad (27)$$

where $\eta = 2\psi(t_1)$ and c_7 is a numerical constant. On the other hand, (4.30) gives

$$\zeta_k(1+\alpha) \leq 3^{n+2}\sqrt{|D|}\alpha^{-1} \quad \text{for} \quad 0 < \alpha \leq \frac{1}{2} . \qquad (28)$$

Without loss of generality, we assume that $2\psi(t) < \frac{1}{2}$ for each t in $\mathbb{R}$. Estimate (19) follows from (20) - (23) combined with (27) and (28). To prove (18) we remark that

$$\log(3^n|D|\log\varphi(t,\chi)) = O(\log\varphi(t,\chi)),$$

therefore, by (19),

$$M = \mathrm{Re}\ f(s_1(t_1)) = O(\log\varphi(t_1,\chi))$$

and (18) follows from (24) and (22). Let $c_4 = (6c_1)^{-1}$, then the region (17) is contained in the union of the circles $|s-s(t_1)| \leq r(t_1)$ and the assertion follows.

Let us recall that $a(\chi) = |D|\cdot|\mathfrak{f}(\chi)|$ and define

$$b(\chi) := \prod_{p\in S_1} (3+|t_p(\chi)|) \prod_{p\in S_2} (3+\frac{|a_p(\chi)|+|t_p(\chi)|}{2})^2. \qquad (29)$$

By (29) and the definition of $\varphi(t,\chi)$, we have

$$\varphi(t,\chi) \leq (1+|t|)^{n/2} \sqrt{a(\chi)b(\chi)}. \qquad (30)$$

<u>Theorem 1</u>. Let $\chi \in gr(k)$. Then

$$\sum_{|p|<x} \chi(p) = g(\chi) \int_2^x \frac{du}{\log u} + O(x^{\alpha}) + O(x \exp(-c_8 \frac{\log x}{\sqrt{n \log x} + \log(a(\chi)b(\chi))}))$$

with $c_8 > 0$, where p ranges over prime ideals of k. Here α denotes the possible exceptional zero of $L(s,\chi)$ in the region defined by (2).

<u>Proof</u>. Let, for $T \geq 1$,

$$\mathcal{L}_T = \{\sigma+it \mid \sigma=1-\psi(t),\ |t| \leq T,\ t \in \mathbb{R}\} \cup$$

$$\{\sigma\pm iT \mid 1+\psi(T) \geq \sigma \geq 1-\psi(T)\},$$

where $\psi(t) = c_4 \log \varphi(t,\chi)$. Since

$$-\frac{L'}{L}(s,\chi) = \sum_{m=1}^{\infty} \sum_{p\in S_o} \chi(p^m) \frac{\log|p|}{|p|^{ms}} \quad \text{for } \operatorname{Re} s > 1,$$

it follows from lemma 4.5 that

$$\sum_{|p|<x} \chi(p) \log|p| = \frac{1}{2\pi i} \int_{1+\psi(T)-iT}^{1+\psi(T)+iT} (-\frac{L'}{L}(s,\chi))\frac{x^s}{s} ds + R(x,T), \qquad (31)$$

where

$$R(x,T) = O(\sum_{m=2}^{\infty} \sum_{|p^m|<x} \log|p|) + O(\frac{x^{1+\psi(T)}}{T\psi(T)}) + O(\frac{nx \log^2 x}{T}). \tag{32}$$

Since, by Lemma 1, $f(s)$ is regular in the region defined by (17), it follows from (31) and (15) that

$$\sum_{|p|<x} \chi(p) \log|p| = g(\chi)x + g_1(\chi) \frac{x^\alpha}{\alpha} + R_1(x,T), \tag{33}$$

where

$$R_1(x,T) = R(x,T) + O(|\int_{\mathcal{L}_T} \frac{x^s}{s} ds \frac{L'}{L}(s,\chi)|). \tag{34}$$

Without loss of generality we may assume that

$$\frac{1}{s-\alpha} = O(\psi(T)^{-1}) \quad \text{for} \quad s \in \mathcal{L}_T. \tag{35}$$

Relations (15), (18) and (35) give:

$$\int_{\mathcal{L}_T} \frac{x^s}{s} ds \frac{L'}{L}(s,\chi) = O(x^{1-\psi(T)}(\log T)\psi(T)^{-2}+x^{1+\psi(T)}(T\psi(T))^{-1}). \tag{36}$$

We remark also that

$$\sum_{m=2}^{\infty} \sum_{|p^m|<x} \log|p| = O(nx^{1/2} \log x) \tag{37}$$

and choose T to satisfy the equation

$$\log x = \frac{\log T}{\psi(T)} . \tag{38}$$

Estimates (32), (34), (36) - (38) combined with (30) give:

$$R_1(x,T) = O(x \exp(-c_9 \frac{\log x}{\log(a(\chi)b(\chi))+\sqrt{n \log x}})), \quad c_9 > 0. \qquad (39)$$

Making use of the partial summation (cf., e.g., [78], p. 371, Satz 1.4 with $A(x) = \sum_{|p|<x} \chi(p) \log|p|$ and $g(\xi) = (\log \xi)^{-1}$) we deduce the assertion of the theorem from (33) and (39).

Let $\rho \in R(k), \chi = \operatorname{tr} \rho$, and suppose that (3.26) holds; we denote by $\alpha(\chi_j)$ the possible exceptional zero of $L(s,\chi_j)$, $1 \le j \le m$, and let

$$n_j = [k_j:\mathbb{Q}], \quad n(\chi) = \sum_{j=1}^{m} n_j. \qquad (40)$$

<u>Theorem 2</u>. Let $\rho \in R(k)$, $\chi = \operatorname{tr} \rho$, then, in notations of (3.26) and (40), we have

$$\sum_{|p|<x} \chi(p) = g(\chi) \int_2^x \frac{du}{\log u} + O(n(\chi)\sqrt{x}) + O(\sum_{j=1}^{m} x^{\alpha(\chi_j)})$$

$$+ O(x \sum_{j=1}^{m} \exp(-c_8 \frac{\log x}{\sqrt{n_j \log x} + \log(a(\chi_j)b(\chi_j))})), \qquad (41)$$

where $g(\chi)$ denotes the multiplicity of the identical representation in ρ and $\chi(p)$ denotes the trace of the operator

$$\rho(\sigma_p): V_p \to V_p.$$

<u>Proof</u>. Equating the Euler factors in (3.26) one obtains an identity:

$$\det(1-\rho(\sigma_p)t)^{-1} = \prod_{j=1}^{m} \prod_{\mathfrak{p}|p} (1-\chi_j(\mathfrak{p})t^{f(\mathfrak{p})})^{-1}, \qquad (42)$$

where p ranges over $S_o(k)$, $\mathfrak{p} \in S_o(k_j)$, $N_{k_j/k}\mathfrak{p} = p^{f(\mathfrak{p})}$. Taking the logarithms in the both sides of (42) we get:

$$\sum_{\nu=1}^{\infty} \frac{t^{\nu}}{\nu} \operatorname{tr}(\rho(\sigma_p)^{\nu}) = \sum_{j=1}^{m} \sum_{\mathfrak{p}|p} \sum_{\nu=1}^{\infty} \frac{t^{\nu f(\mathfrak{p})}}{\nu} e_j \chi_j(\mathfrak{p}^{\nu}). \tag{43}$$

Equating the first terms in (43) we obtain an equation:

$$\sum_{|p|<x} \chi(p) = \sum_{j=1}^{m} e_j \sum_{|\mathfrak{p}|<x}' \chi_j(\mathfrak{p}), \tag{44}$$

where the summation in Σ' is extended over the prime divisors $\mathfrak{p}$ in $S_o(k_j)$ with $f(\mathfrak{p}) = 1$. On the other hand,

$$\sum_{|\mathfrak{p}|<x}' \chi_j(\mathfrak{p}) = \sum_{|\mathfrak{p}|<x} \chi_j(\mathfrak{p}) + O(n_j\sqrt{x}). \tag{45}$$

We remark also that, by (4.32),

$$\sum_{j=1}^{m} e_j g(\chi_j) = g(\chi). \tag{46}$$

Estimate (41) follows now from theorem 1 applied to each of the characters χ_j in (44) when one takes into account relations (45) and (46).

§6. Consequences of the Riemann Hypothesis.

Under assumption (3.32) the estimates obtained in §4 and §5 may be considerably improved. We start with a few general remarks. It follows from (4.17), (5.29) and (5.30) that

$$|L(s,\chi)| \leq 12^n (3|\frac{1+s}{1-s}|^{g(\chi)})(1+|\operatorname{Im} s|)^{n/2} \sqrt{a(\chi)b(\chi)} \tag{1}$$

for $\operatorname{Re} s \geq \frac{1}{2}$, $\chi \in gr(k)$.

Lemma 1. Let $f(s)$ be a function holomorphic in the half-plane $\operatorname{Re} s > \frac{1}{2}$ and suppose that

$$f(s) \neq 0 \quad \text{for} \quad \operatorname{Re} s > \frac{1}{2} ,$$

that

$$|f(s)| < B(f)(1+|t|)^{\ell} \quad \text{with} \quad \ell > 0, \quad B(f) > 1, \tag{2}$$

where $t := \operatorname{Im} s$, and that

$$\log f(s) = O(\ell \log(2+\eta^{-1})) \quad \text{for} \quad \operatorname{Re} s > 1+\eta, \quad \eta > 0. \tag{3}$$

Then

$$\log f(s) = O_{\varepsilon}(\ell[\log(B(f)(1+|t|)^{\ell})]^{2(1-\sigma)+\varepsilon}), \quad \varepsilon > 0, \tag{4}$$

for any σ in the interval $1/2 < \sigma \leq 1$, $\sigma := \operatorname{Re} s$.

Proof. Write $s = \sigma+it$ with $t \in \mathbb{R}$, $1/2 < \sigma \leq 1$. Let

$$\sigma_o(t) = 1+c_1 \log(B(f)(1+|t|)^{\ell}), \quad \delta(t) = \sigma_o(t)^{-1}, \quad c_1 > 0,$$

and suppose that c_1 is chosen to satisfy the condition:

$$\frac{1}{2} + \delta(0) < \sigma \quad \text{and therefore} \quad \frac{1}{2} + \delta(t) < \sigma \quad \text{for any } t.$$

Let

$$s_o = \sigma_o(t)+it,\ r_1 = \sigma_o(t)-(1+\delta(t)),\ r_2 = \sigma_o(t),\ r_3 = \sigma_o(t)-(\tfrac{1}{2}+\delta(t)),$$

and let

$$M_i = \max|g(s)| \quad \text{on the circle} \quad |s-s_o| = r_i, \quad i = 1,2,3,$$

where $g(s) := \log f(s)$. Since, by condition, the function $g(s)$ is holomorphic for $\operatorname{Re} s > \frac{1}{2}$, it follows from the Hadamard's three circles theorem (cf., e.g., [78], p. 401, Satz 9.2) that

$$M_2 \leq M_1^{1-a} M_3^a , \quad a := \left(\log \frac{r_2}{r_1}\right)\left(\log \frac{r_3}{r_1}\right)^{-1} . \tag{5}$$

By (3),

$$M_1 = O(\ell \log(2+\sigma_o(t))). \tag{6}$$

Since

$$0 < a \leq 2(1-\sigma) + 2\delta(t) + 2(4\sigma_o(t)-1)^{-1} ,$$

it follows from (5) and (6) that

$$M_2 = O(\ell(\log \sigma_o(t)) M_3^{2(1-\sigma)+c_2\sigma_o(t)^{-1}}). \tag{7}$$

Let $R = \sigma_o(t) - \frac{1}{2}(1+\delta(t))$ and let

$$M = \max(\mathrm{Re}\ g(s)) \quad \text{on the circle} \quad |s-s_o| = R.$$

By Satz 4.2 in [78, p. 383], we have

$$M_3 \leq g(s_o)(1+\sigma_o(t)^2) + 4M\sigma_o(t)^2. \tag{8}$$

Conditions (2) and (3) give:

$$g(s_o) = O(\ell) \quad \text{and} \quad M = O(\log[B(f)(1+|t|)^{\ell}(1+\sigma_o(t))^{\ell}]),$$

and therefore (4) follows from (8) and (7).

Corollary 1. In notations and under the conditions of lemma 1, we have

$$f(s)^{\alpha} = O_{\varepsilon}((B(f)(1+|t|)^{\ell})^{\ell\varepsilon}) \quad \text{for} \quad \mathrm{Re}\ s \geq \tfrac{1}{2}+\varepsilon, \quad \varepsilon > 0, \tag{9}$$

where $\alpha \in \{\pm 1\}$.

Proof. Write, for brevity, $A = B(f)(1+|t|)$. By (4),

$$|\log f(s)| < C(\sigma_1)\ell(\log A)^{1-\sigma_1} \quad \text{for} \quad \sigma_1 > 0. \tag{10}$$

If A is chosen to be large enough, say $\log A > \left(\frac{C(\sigma_1)}{\sigma_1}\right)^{1/\sigma_1}$, it follows from (10) that

$$|\log f(s)| \leq \ell\sigma_1 \log A,$$

and (9) follows.

Corollary 2. Suppose that $L(s,\chi)$ satisfies (3.32). Then

$$\left[L(s,\chi)\left(\frac{1-s}{1+s}\right)^{g(\chi)}\right]^{\alpha} = O_{\varepsilon}((24^n a(\chi)b(\chi)(1+|t|)^n)^{n\varepsilon}) \tag{L 11}$$

for $\operatorname{Re} s > \frac{1}{2}+\varepsilon$, $\varepsilon > 0$, $\alpha \in \{\pm 1\}$, $t := \operatorname{Im} s$, $\chi \in gr(k)$.

Proof. It follows from (1) and (9) since $L(s,\chi)(\frac{1-s}{1+s})^{g(\chi)}$ is holomorphic for $\operatorname{Re} s > \frac{1}{2}$ and doesn't vanish in this half-plane by assumption.

Relation (11), implied by the Generalised Riemann Hypothesis, may be called the Generalised Lindelöf Hypothesis (cf. [89], Ch. XIII). Let now $\rho \in R(k)$, $\chi = \operatorname{tr} \rho$ and suppose that $L(s,\chi)$ satisfies (3.26); write

$$n_j := [k_j : \mathbb{Q}], \ n(\chi) = \sum_{j=1}^{m} n_j, \quad \tilde{n}(\chi) = \sum_{j=1}^{m} n_j^2 .$$

If each of the functions $L(s,\chi_j)$, $1 \leq j \leq m$, satisfies (11), then

$$[L(s,\chi)(\frac{1-s}{1+s})^{g(\chi)}]^{\alpha} = O_{\varepsilon}(24^{\tilde{n}(\chi)\varepsilon}(1+|t|)^{\tilde{n}(\chi)\varepsilon}(a(\chi)b(\chi))^{n(\chi)\varepsilon}) \qquad \text{(L 12)}$$

for $\operatorname{Re} s > \frac{1}{2}+\varepsilon$, $\varepsilon > 0$, $\alpha \in \{\pm 1\}$, $t := \operatorname{Im} s$.

In particular, (12) follows from the Generalised Riemann Hypothesis.

Remark 1. Alternatively one can deduce an analogous to (12) but a slightly stronger estimate from the two conditions:

i) ρ is of AW type, and

ii) $L(s,\chi) \neq 0$ for $\operatorname{Re} s > \frac{1}{2}$.

Definition 1. If relation (12) holds, we say that ρ is of Lindelöf type.

Proposition 1. Let $\rho \in R(k)$, $\chi = \operatorname{tr} \rho$, and suppose that ρ is of Lindelöf type. Then

$$A(x,\chi) = xP(\chi, \log x) + O(C(\varepsilon,\tilde{n}(\chi))(a(\chi)b(\chi))^{\varepsilon} x^{1/2+\varepsilon}) \qquad \text{(L 13)}$$

where $A(x,\chi)$ and $P(\chi,t)$ have the same meaning as in (4.35), $\varepsilon > 0$ and $C(\varepsilon,\tilde{n}(\chi))$ is a positive constant effectively computable in terms of ε and $\tilde{n}(\chi)$.

Proof. By lemma 4.5 and in view of (4.36), we have

$$A(x,\chi) = \frac{1}{2\pi i}\int_{2-iT}^{2+iT} L(s,\chi)\frac{x^s}{s}\,ds + O\left(\frac{x^2}{T}\right) + O(C_1(\varepsilon,nd(\chi))x^{1+\varepsilon}T^{-1}),$$

where $\varepsilon > 0$ and C_1 is defined as in (4.36). To deduce (13) we move the contour of integration to the line $\mathrm{Re}\ s = \frac{1}{2}+\varepsilon$ and make use of (12). This procedure leads to (13) when one sets $T = x^{3/2}$ and takes into account that, in view of (12), $L(s,\chi)(s-1)^{g(\chi)}$ is regular for $\mathrm{Re}\ s > \frac{1}{2}$.

Lemma 2. Let f be an entire function and suppose that there is a non-decreasing function $\varphi\colon \mathbb{R}_+ \to \mathbb{R}_+$ satisfying the following conditions: $\varphi(u) > 1$ for $u \in \mathbb{R}_+$ and

$$|f(u+it)| < \varphi(|t|) \quad \text{for} \quad a \le u \le b, \quad t \in \mathbb{R},$$

where $a \le -\frac{1}{4}$ and $b \ge 5$. Let

$$N(f,T) := \mathrm{card}\{s \mid f(s) = 0,\ 0 \le \mathrm{Re}\ s \le 1,\ \ 0 \le \mathrm{Im}\ s \le T\}$$

for $T > 0$. Then

$$N(f,T+1) = N(f,T) + O(\log \varphi(T+3)) \tag{14}$$

Proof. Let $\alpha = 2+iT$ and let

$$\nu(u) = \mathrm{card}\{s \mid f(s) = 0,\ |s-\alpha| \le u\}.$$

By a classical theorem (cf. e.g., [88, p. 126], equation (2)),

$$\int_0^{9/4} \frac{\nu(u)}{u} du = \frac{1}{2\pi} \int_0^{2\pi} \log|f(\alpha + \tfrac{9}{4} e^{i\theta})| d\theta - \log|f(\alpha)|. \tag{15}$$

Since the circle $|s-\alpha| = 9/4$ is contained in the strip $a \leq \text{Re } s \leq b$, equation (15) shows that

$$\int_0^{9/4} \frac{\nu(u)}{u} du = O(\log \varphi(T+9/4)). \tag{16}$$

On the other hand,

$$\nu(\sqrt{5}) \log \frac{9}{4\sqrt{5}} = \int_{\sqrt{5}}^{9/4} \nu(\sqrt{5}) \frac{du}{u} \leq \int_0^{9/4} \frac{\nu(u)}{u} du. \tag{17}$$

It follows from (16) and (17) that

$$\nu(\sqrt{5}) = O(\log \varphi(T+3)). \tag{18}$$

But

$$N(f,T+1) \leq N(f,T) + \nu(\sqrt{5}),$$

therefore (14) follows from (18).

Let $\chi \in gr(k)$ and let, for $T > 0$,

$$N(\chi,T) := \text{card}\{s \mid 0 \leq \text{Re } s \leq 1,\ 0 \leq \text{Im } s \leq T,\ L(s,\chi) = 0\}.$$

Proposition 2. Let $\chi \in gr(k)$. Then

$$N(\chi,T+1) = N(\chi,T) + O(\log(a(\chi)b(\chi)(3+T)^n)). \tag{19}$$

Proof. Estimate (19) follows from (4.17) and (5.30) in view of lemma 2.

Lemma 3. Let $\chi \in gr(k)$. Then

$$\frac{L'}{L}(s,\chi) = \sum_{|t-\gamma|<1} (s-\rho)^{-1} - g(\chi)\left(\frac{1}{s}+\frac{1}{s-1}\right) + O(A(\chi,t)), \tag{20}$$

where $-\frac{1}{4} \leq \mathrm{Re}\, s \leq 2$, $t := \mathrm{Im}\, s$, $\gamma := \mathrm{Im}\,\rho$, $A(\chi,t) := \log(a(\chi)b(\chi)\cdot(3+|t|)^n)$ and ρ ranges over the zeros of $L(s,\chi)$ in the strip $0 \leq \mathrm{Re}\, s \leq 1$.

Proof. The function

$$s \mapsto (s(s-1))^{g(\chi)}\Lambda(s,\chi)$$

may be decomposed in a Weierstrass product:

$$(s(s-1)^{g(\chi)}\Lambda(s,\chi) = \exp(A+Bs) \prod_{\rho} \exp\left(\frac{s}{\rho}\right)\left(1-\frac{s}{\rho}\right)$$

(cf., e.g., [78, p. 393], Satz 5.8). Taking the logarithmic derivative in the both sides of this identity one obtains an equation:

$$\frac{L'}{L}(s,\chi) = B + \sum_{\rho}\left(\frac{1}{\rho}+\frac{1}{s-\rho}\right) - \frac{L'_\infty}{L_\infty}(s,\chi) - g(\chi)\left(\frac{1}{s}+\frac{1}{s-1}\right),$$

which leads to the equation:

$$\frac{L'}{L}(s,\chi) = -g(\chi)\left(\frac{1}{s}+\frac{1}{s-1}\right) + \sum_{\rho}\left(\frac{1}{s-\rho} - \frac{1}{2+it-\rho}\right) + R(s), \tag{21}$$

where

$$R(s) = \frac{L'}{L}(2+it,\chi) - \frac{L'_\infty}{L_\infty}(s,\chi) + \frac{L'_\infty}{L_\infty}(2+it,\chi) + g(\chi)\left(\frac{1}{2+it} + \frac{1}{1+it}\right)$$

and $t := \mathrm{Im}\, s$. Since (cf., e.g., [78, p. 395], Satz 6.3)

$$\frac{\Gamma'}{\Gamma}(s) = \log s - \frac{1}{2s} + O_\delta\left(\frac{1}{|s|^2}\right) \quad \text{for } |\arg s| \leq \pi-\delta,\ \delta > 0,$$

and since $\frac{L'_\infty}{L_\infty}(s,\chi)$ is regular in the strip $-\frac{1}{4} \leq \mathrm{Re}\, s \leq 2$, the following estimate holds:

$$R(s) = O(A(\chi,t)) \quad \text{for} \quad -\frac{1}{4} \leq \operatorname{Re} s \leq 2. \tag{22}$$

Let

$$R_1(s) = \sum_{|t-\gamma|\geq 1} \left(\frac{1}{s-\rho} - \frac{1}{2+it-\rho}\right), \quad \gamma := \operatorname{Im} \rho.$$

In view of (19), we have

$$|R_1(s)| \leq 3 \sum_{|t-\gamma|\geq 1} \frac{1}{|\gamma-t|^2} = O(A(\chi,t)) \tag{23}$$

for $-\frac{1}{4} \leq \operatorname{Re} s \leq 2$. Since $|2+it-\rho| \geq 1$ for each ρ, estimate (20) follows from (21) - (23).

<u>Proposition 3</u>. Let $\chi \in gr(k)$ and suppose that $L(s,\chi) \neq 0$ for $\operatorname{Re} s > \frac{1}{2}$. Then

$$\sum_{|p|<x} \chi(p) = g(\chi) \int_2^x \frac{du}{\log u} + O(x^{1/2}(\log(a(\chi)b(\chi))+n \log x)), \tag{R 24}$$

where p ranges over prime ideals in k.

<u>Proof</u>. Let

$$-\frac{L'}{L}(s,\chi) = \sum_{m=1}^{\infty} \lambda(m,\chi)m^{-s}, \qquad \operatorname{Re} s > 1, \tag{25}$$

so that

$$\lambda(m,\chi) = \sum_{p\in S_o(k)}{}' \chi(p^{\ell})\log|p|, \tag{26}$$

where Σ' is a finite sum extended over primes subject to the condition: $|p|^{\ell} = m$, $\ell \geq 1$, $\ell \in \mathbb{Z}$ (in particular, $\lambda(m,\chi) = 0$ when

$m \notin \{|p|^{\ell} | \ell \in \mathbb{Z}, \ell \geq 1, p \in S_o(k)\}$. Since $|\lambda(m,\chi)| \leq n \log m$, it follows from lemma 4.5 that

$$\sum_{m<x} \lambda(m,\chi) = \frac{1}{2\pi i} \int_{\alpha-iT}^{\alpha+iT} \left(-\frac{L'}{L}(s,\chi)\frac{x^s}{s}\right) ds + O\left(\frac{nx(\log x)^2}{T}\right),$$

where $\alpha := 1 + (\log x)^{-1}$, $T \geq 1$. Relation (19) shows that T may be adjusted in such a way that

$$|T-\gamma|^{-1} = O(A(\chi,T))$$

for each ρ in (20). Therefore after moving the line of integration to $\mathrm{Re}\ s = -\frac{1}{4}$ we deduce from (20) the following relation:

$$\sum_{m<x} \lambda(m,\chi) = g(\chi)x + \sum_{|\mathrm{Im}\ \rho|<T} \frac{x^{\rho}}{\rho} + O\left(\frac{x}{T}(n(\log x)^2 + A(\chi,T)^2)\right), \qquad (27)$$

where ρ varies over the zeros of $L(s,\chi)$ in the critical strip $0 \leq \mathrm{Re}\ s \leq 1$ subject to the condition $|\mathrm{Im}\ \rho| < T$. Under the assumption (3.32) it follows from (27) and (19) that

$$\sum_{m<x} \lambda(m,\chi) = g(\chi)x + O(nx^{1/2}\log^2 x) + O(x^{1/2}(\log x)\log(a(\chi)b(\chi))), \qquad \text{(R 28)}$$

if we choose $T = x$. Obviously,

$$\sum_{|p|<x} \chi(p) = \sum_{m<x} \frac{\lambda(m,\chi)}{m} + O(nx^{1/2}). \qquad (29)$$

Relation (24) follows from (28) and (29) by partial summation (cf., e.g., [78, p. 371], Satz 1.4).

<u>Theorem 1</u>. Let $\psi \in R(k)$, $\chi = \mathrm{tr}\ \psi$ and suppose that $L(s,\chi)$ satisfies (3.26) and that $L(s,\chi_j) \neq 0$ for $\mathrm{Re}\ s > \frac{1}{2}$, $1 \leq j \leq m$. Then

$$\sum_{|p|<x} \chi(p) = g(\chi) \int_2^x \frac{du}{\log u} + O(x^{1/2}(\log(a(\chi)b(\chi)) + n(\chi)\log x)), \quad \text{(R 30)}$$

where $a(\chi) = \prod_{j=1}^{m} a(\chi_j)$, $b(\chi) = \prod_{j=1}^{m} b(\chi_j)$, $n(\chi) = \sum_{j=1}^{m} [k_j:\mathbb{Q}]$.

<u>Proof</u>. In view of (5.44) and (5.45), we have

$$\sum_{|p|<x} \chi(p) = \sum_{\substack{|\mathfrak{p}|<x \\ 1\leq j\leq m}} e_j\chi_j(\mathfrak{p}) + O(n(\chi)\sqrt{x}), \quad (31)$$

where $\mathfrak{p}$ ranges over the prime divisors in k_j. Estimate (30) follows from Proposition 3 applied to each of the characters χ_j and relations (31), (5.46).

§7. Equidistribution problems.

Let M be a non-empty set equipped with a (finitely additive positive) measure μ and let E be a system of μ-measurable subsets of M. A subset V of M is said to be n-smooth if there are a positive constant $C(V)$ and, for every Δ in the interval $0 < \Delta < 1$, a system $E_o(\Delta)$ satisfying the following conditions:

1) $E_o(\Delta) \subseteq E$;

2) $\operatorname{card} E_o(\Delta) \leq \Delta^{-n}$;

3) $V \subseteq \bigcup_{\rho \in E_o(\Delta)} \rho$;

4) if $\rho \in E_o(\Delta)$, $\rho' \in E_o(\Delta)$ and $\rho \neq \rho'$, then $\rho \cap \rho' = \emptyset$;

5) $\Big(\bigcup_{\substack{\rho \cap V \neq \emptyset \\ \rho \in E_o(\Delta)}} \rho \setminus \bigcup_{\substack{\rho \subseteq V \\ \rho \in E_o(\Delta)}} \rho \Big) < C(V)\Delta.$

Suppose we are given an infinite set J and two maps

$$f_1 : J \to M \ , \quad f_2 : J \to R_+ \ .$$

The triple (J, f_1, f_2) is said to be (E,μ)-<u>equidistributed</u> if for each ρ in E we have an asymptotic relation of the following shape:

$$\operatorname{card} \{ s \mid s \in J,\ f_1(s) \in \rho,\ f_2(s) < x \} = \mu(\rho) a(x) + O(b(x)), \tag{1}$$

where $a: \mathbb{R}_+ \to \mathbb{R}_+$, $b: \mathbb{R}_+ \to \mathbb{R}_+$ are two monotonely increasing functions such that

$$a(x) \geq b(x) \quad \text{for} \quad x \in \mathbb{R}_+ \quad \text{and} \quad \lim_{x \to \infty} \frac{a(x)}{b(x)} = \infty. \tag{2}$$

<u>Proposition 1</u>. If (J, f_1, f_2) is (E,μ)-equidistributed, then

$$\text{card}\{s \mid s \in J,\ f_1(s) \in V,\ f_2(s) < x\} = \mu(V)a(x) + O(b_1(x)) \tag{3}$$

with $b_1(x) := (C(V)a(x))^{n/n+1} b(x)^{1/n+1}$ for any n-smooth subset V.

Proof. Let $0 < \Delta < 1$, and let

$$V_1 = \bigcup_{\substack{\rho \cap V \neq \emptyset \\ \rho \in E_o(\Delta)}} \rho \quad , \qquad V_2 = \bigcup_{\substack{\rho \subseteq V \\ \rho \in E_o(\Delta)}} \rho \quad .$$

Let

$$\text{card}\{s \mid s \in J,\ f_1(s) \in V,\ f_2(s) < x\} =: \mathcal{N}(V,x).$$

Obviously,

$$\mathcal{N}(V_1,x) \geq \mathcal{N}(V,x) \geq \mathcal{N}(V_2,x) \tag{4}$$

It follows from (1) and (2) that

$$\mathcal{N}(V_j,x) = \mu(V_j)a(x) + O(\Delta^{-n}b(x)), \quad j = 1,2. \tag{5}$$

Since $\mu(V_1 \setminus V_2) < C(V)\Delta$ in view of 5), relations (4) and (5) give:

$$\mathcal{N}(V,x) = \mu(V)a(x) + O(\Delta^{-n}b(x)) + O(C(V)\Delta a(x)). \tag{6}$$

Estimate (3) follows from (6) with $\Delta = \left(\frac{b(x)}{C(V)a(x)}\right)^{1/n+1}$.

Definition 1. Given a triple (M,E,μ), we call elements of E elementary sets. The constant $C(V)$ is called the smoothness constant of V.

Let, in particular, $M = \mathcal{T} \times H$, where $\mathcal{T}$ is an n-dimensional real torus and H is a finite group. To define E we fix a system of generators $\{\lambda_j | 1 \leq j \leq n\}$ of the character group $\hat{\mathfrak{T}}$ (isomorphic to $\mathbb{Z}^n$) and let

$$\lambda_j(u) = \exp(2\pi i \varphi_j(u)), \quad 0 \leq \varphi_j(u) < 1, \quad u \in \mathfrak{T}.$$

An elementary set ρ is, by definition, of the shape

$$\rho = \{u | u \in \mathfrak{T}, \ a_j \leq \varphi_j(u) < b_j, \ 1 \leq j \leq n\} \times g, \tag{7}$$

where g is a conjugacy class in H and a_j, b_j are subject to the condition $0 \leq a_j < b_j \leq 1$. Let μ be the Haar measure on M normalized by the condition $\mu(M) = 1$. Given $\chi \in \hat{M}$, we write

$$\chi = \lambda' \prod_{j=1}^{n} \lambda_j^{\ell_j}, \quad \ell_j \in \mathbb{Z}, \ \lambda' \in \hat{H},$$

and define the weight $w(\chi)$ of χ by the equation

$$w(\chi) = \prod_{j=1}^{n} (1 + |\ell_j|). \tag{8}$$

Let $g(\chi) = 1$ when $\chi = 1$ and $g(\chi) = 0$ when $\chi \neq 1$. Suppose that a triple (J, f_1, f_2) satisfies the following condition:

$$\sum_{s \in J(x)} \chi(f_1(s)) = g(\chi) a(x) + O(b(x, w(\chi))), \quad \chi \in \hat{M}, \tag{9}$$

where $b(x, \ell) \geq 0$, and

$$J(x) = \{s | s \in J, \ f_2(s) < x\}, \quad x > 0.$$ Let

$$b_k(x) = \sum_{\ell=1}^{\infty} d_n(\ell) b(x, \ell) \ell^{-k}, \qquad k \geq 1, \tag{10}$$

where

$$d_n(\ell) = \text{card}\{(\ell_1,\dots,\ell_n) \mid \ell_j \in \mathbb{Z},\ \ell_j \geq 1,\ \prod_{j=1}^{n} \ell_j = \ell\}. \tag{11}$$

Proposition 2. Suppose that both $a(x)$ and $b_k(x)$ are monotonely increasing functions, that $a(x) \geq b_k(x)$ for $x > 0$ and that $a(x)b_k(x)^{-1} \to \infty$ for some k. Then the triple (J,f_1,f_2) is (E,μ)-equidistributed.

The following lemma is elementary.

Lemma 1. Suppose that $0 < \delta < a < b < b+\delta < 1$. Then there is an infinitely differentiable function φ: $[0,1] \to [0,1]$ satisfying the following conditions:

$$\varphi(x) = 1 \quad \text{when} \quad x \in [a,b], \quad \varphi(x) = 0 \quad \text{when} \quad x \notin [a-\delta,b+\delta],$$

$$\varphi(x) = \sum_{\ell=-\infty}^{\infty} c(\ell)\exp(2\pi i \ell x) \quad \text{with} \quad c(0) = b-a+O(\delta),\ c(\ell) = O((\ell\delta)^{-k})$$

for any positive k (the implied by the O-symbol constant may depend on k but not on a,b,δ).

Proof. Let $f \in C^\infty(\mathbb{R})$ and suppose that

$$f'(x) > 0 \quad \text{for} \quad x \in \mathbb{R},\ f(x) = 0 \quad \text{for} \quad x < 0,\ f(x) = 1 \quad \text{for} \quad x > 1$$

(one can, for instance, define f as follows:

$$f(x) = 0 \quad \text{when} \quad x < 0,\ f(x) = 1 \quad \text{when} \quad x > 1 \quad \text{and}$$

$$f(x) = \int_0^u \frac{du}{\exp(u^2(u-1)^2)} \left(\int_0^1 \frac{du}{\exp(u^2(u-1)^2)} \right)^{-1}, \qquad x \in [0,1]).$$

We define φ by the equations:

$$\varphi(x) = f\left(\frac{x-a+\delta}{\delta}\right) \quad \text{when} \quad 0 \le x \le \frac{a+b}{2},$$

$$\varphi(x) = f\left(\frac{b+\delta-x}{\delta}\right) \quad \text{when} \quad \frac{a+b}{2} \le x \le 1.$$

On letting

$$c(\ell) = \int_0^1 \varphi(x)\exp(-2\pi i \ell x)\,dx$$

one checks easily that

$$c(0) = b-a+2\delta \int_0^1 f(x)\,dx$$

and that

$$|c(\ell)| \le \frac{1}{(\delta|\ell|)^k} \max_{0\le x\le 1} |f^{(k+1)}(x)|$$

for $\ell \neq 0$, $k \ge 1$; by construction, $\varphi \in C^\infty[0,1]$ and $\varphi(x) = \sum_{\ell=-\infty}^{\infty} c(\ell)\exp(2\pi i \ell x)$. This proves the lemma.

Let ρ be an elementary set. Without loss of generality we may assume that ρ is given by (7) and that $a_j > 0$, $b_j < 1$, $1 \le j \le n$. Let

$$0 \le a_j - \delta < a_j < b_j < b_j + \delta < 1, \qquad 1 \le j \le n;\ 2\delta < b_j - a_j.$$

By lemma 1, we can construct infinitely differentiable functions

$$\psi^{\pm}_{\delta,j} : \mathbb{R} \to \mathbb{R}, \qquad 1 \le j \le n,$$

satisfying the following conditions:

$$0 \le \psi^{\pm}_{\delta,j}(x) \le 1, \quad \psi^{\pm}_{\delta,j}(x+1) = \psi^{\pm}_{\delta,j}(x) \quad \text{for} \quad x \in \mathbb{R};$$

$$\psi^{\pm}_{\delta,j}(x) = \sum_{\ell=-\infty}^{\infty} c^{\pm}_j(\ell)\exp(2\pi i\ell x), \quad c^{\pm}_j(0) = b_j - a_j + O(\delta),$$

$$c^{\pm}_j(\ell) = O((\ell\delta)^{-k}) \quad \text{for } k > 0, \quad x \in \mathbb{R} ;$$

$$\psi^{+}_{\delta,j}(x) = 1 \quad \text{when} \quad a_j \le x \le b_j,$$

$$\psi^{+}_{\delta,j}(x) = 0 \quad \text{when} \quad x \in [0,1] \quad \text{and} \quad x \notin [a_j-\delta, b_j+\delta],$$

$$\psi^{-}_{\delta,j}(x) = 1 \quad \text{when} \quad x \in [a_j+\delta, b_j-\delta],$$

$$\psi^{-}_{\delta,j}(x) = 0 \quad \text{when} \quad x \in [0,1] \text{ and } x \notin [a_j, b_j].$$

Let $\psi\colon M \to \{0,1\}$ be the characteristic function of ρ in M, i.e. $\psi(x) = 1$ if and only if $x \in \rho$. Orthogonality relations (2.15) show that

$$\psi^{-}_{\delta}(u) \le \psi(u) \le \psi^{+}_{\delta}(u) \quad \text{for} \quad u \in M, \tag{12}$$

where

$$\psi^{\pm}_{\delta}(u) = \frac{|g|}{|H|} \sum_{\chi \in \hat{H}} \overline{\chi(g)}\chi(u) \prod_{j=1}^{n} \psi^{\pm}_{\delta,j}(\varphi_j(u)), \qquad u \in M. \tag{13}$$

Obviously,

$$\text{card}\{s \mid s \in J,\ f_1(s) \in \rho,\ f_2(s) < x\} = \sum_{s\in J(x)} \psi(f_1(s)). \tag{14}$$

On the other hand, by (13),

$$\psi^{\pm}_{\delta}(u) = \frac{|g|}{|H|} \sum_{\chi,\ell} \overline{\chi(g)}\ c^{\pm}(\vec{\ell})\chi(u)\lambda^{\vec{\ell}}(u), \qquad u \in M, \tag{15}$$

where χ ranges over $\hat{H}$, $\vec{\ell} = (\ell_1,\dots,\ell_n)$ ranges over $\mathbb{Z}^n$, and

$$\lambda^{\vec{\ell}}(u) := \prod_{j=1}^{n} (\lambda_j(u))^{\ell_j} \quad \text{with} \quad \lambda_j(u) = \exp(2\pi i \varphi_j(u)),$$

$$c^{\pm}(\vec{\ell}) := \prod_{j=1}^{n} c_j^{\pm}(\ell_j).$$

Relations (9) and (15) give:

$$\sum_{s\in J(x)} \psi_\delta^{\pm}(f_1(s)) = \frac{|g|}{|H|} c^{\pm}(0)a(x)+O(|g| \sum_{\vec{\ell}} |c^{\pm}(\vec{\ell})| b(x, \|\vec{\ell}\|)), \tag{16}$$

where $\|\vec{\ell}\| := \prod_{j=1}^{n} (1+|\ell_j|)$. The properties of the coefficients $c_j^{\pm}(\ell_j)$ and relation (16) lead to the following estimate:

$$\sum_{s\in J(x)} \psi_\delta^{\pm}(f_1(s)) = \mu(\rho)a(x)+O(\delta a(x))+O(|g|\delta^{-kn} b_k(x)).$$

This estimate, combined with (12), gives:

$$\sum_{s\in J(x)} \psi(f_1(s)) = \mu(\rho)a(x)+O(|H| a_1(x)) \tag{17}$$

with

$$a_1(x) = a(x)^{kn/1+kn} \, b_k(x)^{1/1+kn} \tag{18}$$

as soon as one takes $\delta = (b_k(x))^{1/1+kn}(a(x))^{-1/1+kn}$. Since $b_k(x)a(x)^{-1} \underset{x\to\infty}{\to} 0$ for some k, it follows from (17), (18) and (14) that the triple (J,f_1,f_2) is (E,μ)-equidistributed. This proves proposition 2.

<u>Corollary 1</u>. In notations and under the conditions of proposition 2, we have

$$\text{card}\{s | s \in J(x),\ f_1(s) \in \tau\} = \mu(\tau)a(x)+O(b(s)) \tag{19}$$

with

$$b(x) := C(\tau)^{\frac{n}{n+1}} \, |H|\, a(x)^{1-(n+1)^{-1}(kn+1)^{-1}} b_k(x)^{(n+1)^{-1}(kn+1)^{-1}}$$

for any n-smooth subset τ; $k \geq 1$.

Proof. Estimate (19) follows from (14), (17) and proposition 1.

Let now $J = I_o(k)$, $H = H(\tilde{\mathfrak{m}})$, and let $\mathcal{T}$ be defined by (1.3). Corollary 1 and Theorem 4.1 lead to the following assertion.

Theorem 1. In notations of §1,

$$\iota(x;A,\tau) = \frac{w(k)}{h\varphi(\tilde{\mathfrak{m}})} \mu(\tau)x + O(h\varphi(\tilde{\mathfrak{m}})C(\tau)C_o(k)x^{1-\gamma(n)}), \quad \gamma(n) > 0, \qquad (20)$$

for any smooth subset τ of $\mathcal{T}$. Here $C_o(k)$ and $\gamma(n)$ are exactly computable constants depending on the field k and its degree n, respectively.

Secondly, let $J = S_o(k)$, let $H = G(K|k)$ for a finite Galois extension $K|k$, and let $\mathcal{T}$ be the torus defined by (1.3). We denote by $(\frac{\cdot}{K|k})$ the Artin symbol corresponding to the extension $K|k$ and recall that, for $p \in S_o(k)$,

$$\left(\frac{p}{K|k}\right) = \{\sigma \mid \sigma \in G(K|k),\ \alpha^{\sigma} \equiv \alpha^{|p|}\ (\mathfrak{p}) \ \text{ for each } \alpha \text{ in } v_K\},$$

where $\mathfrak{p}|p$, $\mathfrak{p} \in S_o(K)$. Thus if p does not ramify in $K|k$ Artin symbol defines a conjugacy class in $G(K|k)$. Let

$$\pi(x;g,\tau) = \mathrm{card}\{p \mid p \in J(x)\ f(p) \in \tau,\ p \nmid D_{K|k}, \left(\frac{p}{K|k}\right) = g\},$$

where f is defined by (1.4), g is a conjugacy class in $G(K|k)$ and $D_{K|k}$ is the relative discriminant of $K|k$.

Theorem 2. Let τ be a smooth subset of $\mathcal{T}$. Then

$$\pi(x; g, \tau) = \frac{|g|\mu(\tau)}{[K:k]} \int_2^x \frac{du}{\log u} + R(x, K), \tag{21}$$

where

$$R(x, K) = O(C(\tau)x \exp(-C_1(K)\sqrt{\log x})), \quad C_1(K) > 0. \tag{22}$$

Moreover, assuming the Generalised Riemann Hypothesis (3.32) one obtains an estimate

$$R(x, K) = O(C(\tau)C_2(K)x^{1-\gamma_1(n)}), \quad \gamma_1(n) > 0. \tag{R 23}$$

Proof. The assertion of theorem 2 follows from (19) combined with theorem 5.2 and theorem 6.1. Estimates (5.41), (6.30) and (19) allow for effective evaluation of $C_1(K)$, $C_2(K)$ and $\gamma_1(n)$.

We are indebted to Professor W. Narkiewicz for an important remark relating to Theorem 2.

Appendix 1. Frobenius classes in Weil's groups.

Let $K|k$ be a finite Galois extension and let $G := W_1(K|k)$. Consider the set

$$\Sigma = \bigcup_{p \in S_o(k)} \bigcup_{h \in G} h^{-1} \sigma_p h$$

of all the elements in G which lie in one of the conjugacy classes $\{g\}$, $g \in \sigma_p$, $p \in S_o(k)$. The following proposition is a generalisation of a classical theorem asserting that each conjugacy class in the Galois group $G(K|k)$ contains a Frobenius automorphism.

Proposition 1. The set Σ is everywhere dense in G.

Proof. Suppose, on the contrary, that there is an open set V for which

$$\Sigma \cap V = \emptyset \tag{1}$$

and let $g \in V$. By a classical lemma (cf., e.g., [77], §14B) and §16F)), there is a continuous function

$$f: G \to [0,1]$$

such that

$$f(g) = 1, \quad f(h) = 0 \quad \text{when } h \notin V. \tag{2}$$

Since $h^{-1}\Sigma h = \Sigma$ for each h in G, we may assume, without loss of generality, that

$$h^{-1}Vh = V \quad \text{for } h \in G. \tag{3}$$

Let

$$f_1(h) = [\int_G d\mu(u) f(u^{-1}hu)][\int_G d\mu(u) f(u^{-1}gu)]^{-1}, \quad h \in G. \tag{4}$$

It follows from (2) - (4) that

$$f_1(g) = 1, \quad f_1(h) = 0 \text{ when } h \notin V, \quad f_1(h) \geq 0 \text{ for } h \in G, \tag{5}$$

and that

$$f_1(u^{-1}hu) = f_1(h) \quad \text{for} \quad u \in G, \quad h \in G. \tag{6}$$

Since G is a compact group and since $f_1 \colon G \to \mathbb{R}$ is a continuous function, it follows from (6) that f_1 can be decomposed in a Fourier series

$$f_1 = \sum_{\chi \in \hat{G}} a(f_1, \chi)\chi \tag{7}$$

and that this series is uniformly convergent on G. By (1), (5) and (7),

$$\sum_{\chi \in \hat{G}} a(f_1, \chi)\chi(\sigma_p) = 0 \quad \text{for} \quad p \in S_o(k). \tag{8}$$

On the other hand, the prime number theorem (5.41) gives:

$$\sum_{|p|<x} \chi(\sigma_p) = g(\chi) \int_2^x \frac{du}{\log u} + O(x \exp(-c(\chi)\sqrt{\log x})) \tag{9}$$

with $c(\chi) > 0$. Let $\varepsilon > 0$ and let $\mathfrak{N}_\varepsilon$ be a <u>finite</u> subset of $\hat{G}$ such that

$$|f_1(h) - \sum_{\chi \in \mathfrak{N}_\varepsilon} a(f_1, \chi)\chi(h)| < \varepsilon \quad \text{for} \quad h \in G. \tag{10}$$

It follows from (7), (8) and (10) that

$$\Big|\sum_{\chi\in\mathfrak{N}_\varepsilon} a(f_1,\chi)\chi(\sigma_p)\Big| < \varepsilon \quad \text{for} \quad p \in S_o(k). \tag{11}$$

Let $b(f_1) = \int_G f_1(h)d\mu(h)$. In view of (7), we have

$$|a(f_1,\chi)| \leq b(f_1) \quad \text{for} \quad \chi \in \hat{G} \tag{12}$$

and

$$a(f_1,1) = b(f_1). \tag{13}$$

Since $\mathfrak{N}_\varepsilon$ is a finite set, estimate (9) shows that

$$\Big|\sum_{|p|<x} \chi(\sigma_p) - g(\chi)\int_2^x \frac{du}{\log u}\Big| < \frac{\varepsilon}{2|\mathfrak{N}_\varepsilon|b(f_1)}\int_2^x \frac{du}{\log u} \tag{14}$$

for $\chi \in \mathfrak{N}_\varepsilon$, as soon as $x > N(\varepsilon)$ with large enough $N(\varepsilon)$. If ε is chosen to be small enough (say, $0 < \varepsilon < \frac{1}{2}b(f_1)$), then it follows from (10) that $1 \in \mathfrak{N}_\varepsilon$ and therefore

$$\sum_{\chi\in\mathfrak{N}_\varepsilon} g(\chi)a(f_1,\chi) = a(f_1,1). \tag{15}$$

Relations (13) - (15) give:

$$\Big|\sum_{\substack{\chi\in\mathfrak{N}_\varepsilon \\ |p|<x}} a(f_1,\chi)\chi(\sigma_p) - b(f_1)\int_2^x \frac{du}{\log u}\Big| < \frac{\varepsilon}{2}\int_2^x \frac{du}{\log u} \tag{16}$$

for $x > N(\varepsilon)$. If $0 < \varepsilon < \frac{1}{2}b(f_1)$ relations (16) and (11) lead to a contradiction, therefore (1) is impossible.

<u>Corollary 1</u>. Let $\rho \in R(k)$, $\rho' \in R(k)$, $\chi = \operatorname{tr}\rho$, $\chi' = \operatorname{tr}\rho'$ and suppose $\chi(\sigma_p) = \chi'(\sigma_p)$ for each p in $S_o(k)$. Then representations ρ and ρ' are equivalent.

<u>Proof</u>. It follows from proposition 1 and lemma 2.1.

Appendix 2. Ideal classes and norm-forms.

We recall here the definition of "ideal numbers" introduced by E. Hecke and the classical relation between ideal classes and norm-forms. Let, in notations of (1.3),

$$\mathcal{T}_1 = (\mathbb{Z}/2\mathbb{Z})^{r_0} \times \mathcal{T}$$

and let

$$\pi: X^* \to \mathcal{T}_1$$

be the natural projection of X^* on $\mathcal{T}_1$. Let

$$Nx = \prod_{p \in S_\infty} \|x_p\|_p \quad \text{for} \quad x = \prod_{p \in S_\infty} x_p, \quad x \in X,$$

and consider a group

$$W = \{x \mid x \in X^*, \quad Nx = 1\}.$$

By construction, the sequence of groups

$$1 \to V^* \to W \overset{\pi}{\to} \mathcal{T}_1 \to 1$$

is exact. A subset U of W is said to be toroidal if π separates points on U, that is if $\pi(x) \neq \pi(x')$ when $x \neq x'$, $x \in U$, $x' \in U$. Consider the field K obtained by adjoining to k all the roots $\sqrt[h]{\alpha}$, $\alpha \in k$, where h is the class number of k. We chose a $\mathbb{Z}$-basis $A_1, \ldots, A_q$ of H and denote by ℓ_j the order of A_j in H, so that

$$H \cong \mathbb{Z}/\ell_1\mathbb{Z} \oplus \ldots \oplus \mathbb{Z}/\ell_q\mathbb{Z} .$$

Let us fix an integral ideal $\mathfrak{b}_j$ in the ideal class A_j^{-1} and an element β_j in K^* satisfying the following conditions:

$$\beta_j^{\ell_j} \in k, \quad (\beta_j^{\ell_j}) = \mathfrak{b}_j^{\ell_j} \quad , \qquad 1 \le j \le q.$$

For $A = \prod_{j=1}^{q} A_j^{m_j}$ with $0 \le m_j \le \ell_j - 1$, let

$$\mathfrak{b}(A) = \prod_{j=1}^{q} \mathfrak{b}_j^{m_j} \, , \qquad \beta(A) = \prod_{j=1}^{q} \beta_j^{m_j} \qquad .$$

Thus $\mathfrak{b}(A) \in A^{-1}$. Choose a $\mathbb{Z}$-basis $\{w_j(A) | 1 \le j \le n\}$ of the ideal $\mathfrak{b}(A)$ and let

$$f_A(x) = N_{k(x)/\mathbb{Q}(x)}\left(\sum_{j=1}^{n} x_j w_j(A)\right) N_{k/\mathbb{Q}}\, \mathfrak{b}(A)^{-1}, \tag{1}$$

where $x = (x_1, \ldots, x_n)$ is an array of n independent variables. Obviously, $f_A(x) \in \mathbb{Z}[x]$. Moreover, up to unimodular equivalence, the form $f_A(x)$ depends neither on the choice of $\mathfrak{b}(A)$ in A^{-1}, nor on the choice of an integral basis $\{w_j(A)\}$ of $\mathfrak{b}(A)$. Let

$$\lambda_A(a) = \mathfrak{b}(A)^{-1}\left(\sum_{j=1}^{n} a_j w_j(A)\right), \quad a = (a_1, \ldots, a_n) \in \mathbb{Q}^n \, ,$$

so that

$$\lambda_A\colon \mathbb{Q}^n \to A \cup \{0\}$$

maps $\mathbb{Q}^n$ on $A \cup \{0\}$. Moreover,

$$|f_A(a)| = N_{k/\mathbb{Q}}(\lambda_A(a)) \quad \text{for} \quad a \in \mathbb{Q}^n. \tag{2}$$

Obviously, $\lambda_A(a) \in I_0(k)$ if and only if $a \in \mathbb{Z}^n \setminus \{0\}$. We extend the diagonal embedding of k in X to an isomorphism

$$\sigma\colon K \to X,$$

and define a linear operator

$$g_A\colon \mathbb{R}^n \to X$$

by letting

$$g_A(a_1,\dots,a_n) = \sum_{j=1}^{n} a_j\sigma(w_j(A)\beta(A)^{-1}), \quad a_j \in \mathbb{R}.$$

One remarks that

$$|f_A(a)| = N(g_A(a)) \quad \text{for} \quad a \in \mathbb{R}^n. \tag{3}$$

For each $\mathfrak{a}$ in A let us choose α in k so that

$$(\alpha) = \mathfrak{a}\,\mathfrak{b}(A)\;;$$

let

$$\psi(\mathfrak{a}) = \pi(\sigma(\alpha\beta(A)^{-1})). \tag{4}$$

Obviously,

$$\psi(\lambda_A(a)) = \pi(g_A(a)) \quad \text{for} \quad a \in \mathbb{Q}^n\setminus\{0\}. \tag{5}$$

Moreover, it can be easily seen (cf. the proof of Proposition 2 in [65]) that

$$\psi\colon I(k) \to \mathcal{T}_1$$

is a homomorphism. The assignment

$$\mathfrak{a} \mapsto \alpha\beta(A)^{-1}, \quad \mathfrak{a} \in A,$$

may be viewed as a definition of ideal numbers in the sense of E. Hecke.

Let $U \subseteq W$ and let $m \in \mathbb{Z}$, $m > 0$. We define two sets:

$$\mathcal{A}_1(m,U) = \{a \mid a \in \mathbb{Z}^n, |f_A(a)| = m,\ g_A(a)m^{-1/n} \in U\} \tag{6}$$

and

$$\mathcal{A}_2(m,U) = \{\mathfrak{a} \mid \mathfrak{a} \in I_0(k) \cap A,\ N_{k/\mathbb{Q}}\mathfrak{a} = m,\ \psi(\mathfrak{a}) \in \pi(U)\}. \tag{7}$$

Proposition 1. If U is a toroidal subset of W, then λ_A defines a one-to-one correspondence between $\mathcal{A}_1(m,U)$ and $\mathcal{A}_2(m,U)$.

Proof. It is an immediate consequence of the relations (2) and (5) - (7) and the definition of toroidal subsets.

In view of proposition 1, theorem 7.1 and theorem 7.2 may be regarded as statements about equidistribution of integral points on a norm-form variety. To be more precise, let

$$V^{(A)} = \{a \mid a \in \mathbb{R}^n, |f_A(a)| = 1\}.$$

Since $\det g_A \neq 0$ and because of (3), the map

$$g_A: V^{(A)} \to W$$

is a homeomorphism of $V^{(A)}$ on W. Let

$$E = \{U \mid U \subseteq V^{(A)}, \quad g_A(U) \text{ is toroidal, } \pi(g_A(U)) \text{ is smooth}\}.$$

We define now a measure $\mu^{(A)}$ on $V^{(A)}$. Let μ be the Haar measure on $\mathcal{T}_1$ normalised by the condition $\mu(\mathcal{T}_1) = 1$, and let $\tilde{\mu}$ be the v^*-invariant Borel measure on W uniquely defined by the conditions:

$$\tilde{\mu}(U) = \mu(\pi(U)) \quad \text{if } U \text{ is a toroidal subset of } W,$$

and

$$\tilde{\mu}(\sigma(\varepsilon)U) = \tilde{\mu}(U) \quad \text{for } \varepsilon \in v^*, \quad U \subseteq W.$$

Let

$$\mu^{(A)}(U) = \tilde{\mu}(g_A(U)) \quad \text{for } U \subseteq V^{(A)},$$

and let

$$V_o^{(A)} = \{a \mid a \in \mathbb{R}^n, \ f_A(a) = 0\}.$$

One defines two maps

$$h: \mathbb{R}^n \setminus V_o^{(A)} \to V^{(A)}, \quad h: a \mapsto a|f_A(a)|^{-1/n}$$

and

$$h': \mathbb{R}^n \to \mathbb{R}_+ \cup \{0\}, \quad h': a \mapsto |f_A(a)|.$$

The following statement is an immediate consequence of proposition 1, theorem 7.1, theorem 7.2 and the definitions.

Proposition 2. Each of the triples

$$(\lambda_A^{-1}(I_o \cap A),h,h') \quad \text{and} \quad (\lambda_A^{-1}(S_o \cap A),h,h')$$

is $(E,\mu^{(A)})$-equidistributed.

Notations. For $x \in \mathbb{C}^\ell$, $x = (x_1,\ldots,x_\ell)$, we write

$$|x| = \max_{1 \le i \le \ell} |x_i|$$

and let

$$B(y) = \{a \mid a \in V^{(A)}, |g_A(a)| < y\}, \quad y > 0.$$

Lemma 1. Suppose that $n = 2r_2$ and let $r = r_2 - 1$. There is a positive constant $C_1(k)$ depending on r and the regulator of k only and such that

$$\tilde{\mu}(g_A(B(x))) = C_1(k)(\log x)^r \quad \text{for} \quad x > 1. \tag{8}$$

Proof. It is a volume computation: let

$$\alpha_p = \exp(z_p + 2\pi i \varphi_p), \; z_p \in \mathbb{R}, \; 0 \le \varphi_p < 1, \quad p \in S_\infty$$

for $\alpha = \prod_{p \in S_\infty} \alpha_p$, $\alpha \in W$ and let

$$z_p = \sum_{q=1}^{r} \omega_q \log|\sigma_p(\varepsilon_q)|,$$

where $\{\varepsilon_1,\ldots,\varepsilon_r\}$ is a system of fundamental units in k and σ_p: $k \to k_p$ is the natural embedding of k in k_p. By definition,

$$d\tilde{\mu}(\alpha) = d\omega_1 \ldots d\omega_r \, d\varphi_1 \ldots d\varphi_{r+1} .$$

Therefore

$$\tilde{\mu}(g_A(B(x))) = \int_{B_1(x)} d\omega_1 \ldots d\omega_r ,$$

where

$$B_1(x) = \{\omega \mid \sum_{q=1}^{r} \omega_q \log|\sigma_p(\varepsilon_q)| < \log x \quad \text{for} \quad p \in S_\infty\},$$

and (8) follows.

Chapter II. Scalar product of L-functions.

§1. Definition and elementary properties of scalar products.

Given power series

$$P_i(t) = \sum_{n=0}^{\infty} a_n^{(i)} t^n, \quad 1 \le i \le r, \tag{1}$$

and Dirichlet series

$$L_i(s) = \sum_{\mathfrak{a} \in I_o(k)} a_i(\mathfrak{a}) |\mathfrak{a}|^{-s}, \quad 1 \le i \le r, \tag{2}$$

with coefficients $a_n^{(i)}, a_i(\mathfrak{a})$ in a (commutative) field F of characteristic zero, let us define the Hadamard convolution of (1) by the equation

$$(P_1 * \ldots * P_n)(t) = \sum_{n=0}^{\infty} t_n \prod_{i=1}^{r} a_n^{(i)} \tag{3}$$

and the scalar product of (2) by the equation

$$(L_1 * \ldots * L_r)(s) = \sum_{\mathfrak{a} \in I_o(k)} |\mathfrak{a}|^{-s} \prod_{i=1}^{r} a_i(\mathfrak{a}), \tag{4}$$

respectively. Suppose that

$$a_i : I_o(k) \to F, \quad 1 \le i \le r,$$

is a multiplicative function, that is

$$a_i(\mathfrak{a}_1 \mathfrak{a}_2) = a_i(\mathfrak{a}_1) a_i(\mathfrak{a}_2) \quad \text{when} \quad (\mathfrak{a}_1, \mathfrak{a}_2) = 1.$$

Then

$$L_i(s) = \prod_{p \in S_o} \ell_{ip}(|p|^{-s}), \; \ell_{ip}(t) := \sum_{n=0}^{\infty} a_i(p^n) t^n, \quad 1 \le i \le r,$$

and therefore

$$(L_1 * \ldots * L_r)(s) = \prod_{p \in S_o} (\ell_{1p} * \ldots * \ell_{rp})(|p|^{-s}). \tag{5}$$

We start with a few formal lemmas concerning properties of the convolution (3) in the ring $\mathcal{R} = F[[t]]$ of formal power series. Let

$$D: \sum_{n=0}^{\infty} a_n t^n \mapsto \sum_{n=0}^{\infty} (n+1) a_{n+1} t^n$$

be the operator of formal differentiation in $\mathcal{R}$.

<u>Lemma 1</u>. Let $g \in \mathcal{R}$, $\alpha \in F$, $\nu \in \mathbb{N}$. Then

$$(1-\alpha t)^{-(\nu+1)} * g(t) = \frac{1}{\nu!} D^{\nu}(t^{\nu} g(\alpha t)). \tag{6}$$

<u>Proof</u>. For $\nu = 0$ identity (6) follows from the definition (3). One remarks that

$$Df*g = D(f*(tg)) \quad \text{for} \quad f \in \mathcal{R}, \quad g \in \mathcal{R}. \tag{7}$$

Identity (7), with $f(t) = (1-\alpha t)^{-\nu}$, gives:

$$(1-\alpha t)^{-(\nu+1)} * g(t) = \frac{1}{\nu\alpha} D((1-\alpha t)^{-\nu} * (tg(t))) \quad \text{for} \quad \nu \geq 1. \tag{8}$$

Proceeding by induction on ν suppose that (6) holds for $\nu = \mu-1$. Then it follows from (8) that

$$(1-\alpha t)^{-(\mu+1)} * g(t) = \frac{1}{\mu\alpha} D(\frac{1}{(\mu-1)!} D^{\mu-1}(t^{\mu-1}(\alpha t g(\alpha t)))), \tag{9}$$

and therefore we get (6) with $\nu = \mu$. This proves the lemma.

<u>Lemma 2</u>. Let $P(t) \in F[t]$ and $Q(t) \in F[t]$. Then

$$D^{\nu}(t^{\nu}\frac{P(t)}{Q(t)}) = \frac{P_1(t)}{Q_1(t)}, \quad P_1(t) \in F[t], \quad Q_1(t) \in F[t] \quad ,$$

and

$$\deg P_1 - \deg Q_1 \leq \deg P - \deg Q . \tag{10}$$

Proof. For $\nu = 0$ the assertion is obvious. Let $\nu \geq 1$; then

$$D^{\nu}(t^{\nu}P(t)Q(t)^{-1}) = tD^{\nu}(t^{\nu-1}P(t)Q(t)^{-1}) + \nu D^{\nu-1}(t^{\nu-1}P(t)Q(t)^{-1}), \quad (11)$$

by the binomial formula applied to $D^{\nu}(f_1f_2)$ with $f_1(t) = t$ and $f_2(t) = t^{\nu-1}P(t)Q(t)^{-1}$. Proceeding by induction on ν suppose that the assertion of the lemma holds when $\nu = \mu-1$, $\mu \geq 1$. Then it follows from (11) that

$$D^{\mu}(t^{\mu}\frac{P(t)}{Q(t)}) = tD(\frac{P_1(t)}{Q_1(t)}) + \mu\frac{P_1(t)}{Q_1(t)}, \quad P_1(t) \in F[t],\ Q_1(t) \in F[t].$$

Thus

$$D^{\mu}(t^{\mu}\frac{P(t)}{Q(t)}) = \frac{P_2(t)}{Q_2(t)}$$

with

$$P_2(t) = \mu P_1(t)Q_1(t)+t(P_1'(t)Q_1(t)-P_1(t)Q_1'(t)), \quad Q_2(t) = Q_1(t)^2,$$

where we write, for brevity, $f' = Df$. It follows from (10) therefore, that

$$\deg P_2 - \deg Q_2 \leq \deg P - \deg Q .$$

This proves the lemma.

Lemma 3. Let $\alpha_{ij} \in F^*$, $Q_i(t) = \prod_{j=1}^{n_i}(1-\alpha_{ij}t)$ and let $P_i(t) \in F[t]$, $1 \leq i \leq r$. Then

$$\frac{P_1}{Q_1} * \ldots * \frac{P_r}{Q_r} = \frac{U_r}{V_r}, \quad U_r(t) \in F[t],$$

and

$$V_r(t) = \prod_{\alpha\in A}(1-t|\alpha|), \quad A = \{(\alpha_{1j_1},\ldots,\alpha_{rj_r}) \mid {1 \leq j_i \leq n_i \atop 1 \leq i \leq r}\},$$

$|(x_1,\ldots,x_r)| := \prod_{i=1}^{r} x_i$. Moreover, if $\deg P_i < \deg Q_i$ for each i, then

$$\deg U_r \leq \prod_{i=1}^{r} n_i - 1. \quad (12)$$

Proof. Clearly it is enough to prove the assertion in the case $r = 2$.

Suppose that $r = 2$ and let, for brevity,

$$Q_1(t) = \prod_{j=1}^{\nu} (1-\alpha_j t)^{\ell_j}$$

with $\ell_j \geq 1$ and $\alpha_j \neq \alpha_i$ when $i \neq j$. Let

$$\frac{P_1(t)}{Q_1(t)} = \sum_{j=0}^{m} b_j t^j + \sum_{j=1}^{\nu} \sum_{i=0}^{\ell_j-1} b_{ij}(1-\alpha_j t)^{-i-1} \quad . \tag{13}$$

Since $*$ is a linear operation, it follows from (6) and (13) that

$$\frac{P_1(t)}{Q_1(t)} * \frac{P_2(t)}{Q_2(t)} = \left(\sum_{j=0}^{m} b_j t^j\right) * \frac{P_2(t)}{Q_2(t)} + \sum_{j=1}^{\nu} \sum_{i=0}^{\ell_j-1} b_{ij} D^i \left(t^i \frac{P_2(\alpha_j t)}{Q_2(\alpha_j t)}\right) \frac{1}{i!}$$

The first assertion of the lemma follows from this identity. To prove (12) we notice that if $\deg P_1 < \deg Q_1$, then $m = 0$ and $b_o = 0$. Therefore it follows from lemma 2 that

$$\frac{P_1(t)}{Q_1(t)} * \frac{P_2(t)}{Q_2(t)} = \sum_{j=1}^{\nu} \sum_{i=0}^{\ell_j-1} b_{ij} \frac{P_{ij}(t)}{Q_{ij}(t)}$$

with $\deg P_{ij} - \deg Q_{ij} \leq \deg P_2 - \deg Q_2$ for each i,j. Thus, if also $\deg P_2 < \deg Q_2$, we get (12).

<u>Lemma 4</u>. In notations and under assumptions of lemma 3, if

$$P_j(t) = 1 \quad \text{for each } j, \tag{14}$$

then

$$U_r(t) \equiv 1 \pmod{t^2}. \tag{15}$$

<u>Proof</u>. Let

$$\frac{1}{Q_1} * \ldots * \frac{1}{Q_r} = Q.$$

By lemma 3, it follows from (14) that

$$U_r(t) \equiv Q\Big(1-t \sum_{\alpha \in A} |\alpha|\Big) \pmod{t^2}. \tag{16}$$

Since

$$Q_i(t)^{-1} \equiv 1+t \sum_{j=1}^{n_i} \alpha_{ij} \pmod{t^2},$$

we get, by the definition of Q,

$$Q(t) \equiv 1+t \prod_{i=1}^{r} \left(\sum_{j=1}^{n_i} \alpha_{ij} \right) \pmod{t^2}. \tag{17}$$

Comparing (16) and (17) one obtains (15), and lemma follows.

Returning to the identity (5) let us assume, as it is the case, that

$$\ell_{ip}(t) = \prod_{j=1}^{n_i} (1-\alpha_{ij}(p)t)^{-1}, \quad 1 \leq i \leq r, \quad p \in S_o, \tag{18}$$

and let

$$A(p) = \{(\alpha_{1j_1}(p), \ldots, \alpha_{rj_r}(p)) \mid 1 \leq j_i \leq n_i, \ 1 \leq i \leq r\},$$

$$\ell_p(t) = \prod_{\alpha \in A(p)} (1-|\alpha|t)^{-1} . \tag{19}$$

Given a system of polynomials

$$\{\Phi_p(t) \mid p \in S_o, \ \Phi_p(t) \in F[t]\}, \tag{20}$$

let us define a formal Euler product

$$L(s,\Phi) = \prod_{p \in S_o} \Phi_p(|p|^{-s})^{-1} . \tag{21}$$

<u>Proposition 1</u>. Assume (18). Then there is a system of polynomials (20) satisfying the following conditions:

$$\Phi_p(t) \equiv 1 \pmod{t^2}, \quad \deg \Phi_p \leq \prod_{i=1}^{r} n_i - 1, \quad p \in S_o, \tag{22}$$

and

$$(L_1 * \ldots * L_r)(s) = L(s,\Phi)^{-1} \prod_{p \in S_o} \ell_p(|p|^{-s}) \tag{23}$$

in the ring of formal Dirichlet series over k with coefficients in F.

<u>Proof</u>. By lemma 4, it follows from (18) and (19) that

$$\ell_{1p}*\dots*\ell_{rp} = \Phi_p \ell_p \ , \quad p \in S_o, \tag{24}$$

with Φ_p satisfying (22). Identity (23) follows from (24), (21) and (5). This proves the proposition.

Returning to notations of §I.3, let $\rho_j \in R(k)$, $1 \leq j \leq r$, and let

$$L(s,\chi_j) = \sum_{\mathfrak{m}\in I_o(k)} a(\mathfrak{m},\chi_j)|\mathfrak{m}|^{-s} \ , \quad \chi_j = \mathrm{tr}\ \rho_j,\ 1 \leq j \leq r, \tag{25}$$

be the corresponding L-function. We define the scalar product $L(s,\vec{\chi})$ of the L-functions (25) by the identity:

$$L(s,\vec{\chi}) = \sum_{\mathfrak{m}\in I_o(k)} |\mathfrak{m}|^{-s} \prod_{j=1}^{r} a(\mathfrak{m},\chi_j), \quad \vec{\chi} := (\chi_1,\dots,\chi_r). \tag{26}$$

Let

$$\ell_p(\vec{\chi},t)^{-1} = \ell_p(\rho_1,t)^{-1}*\dots*\ell_p(\rho_r,t)^{-1}. \tag{27}$$

<u>Corollary 1</u>. There is a system of polynomials

$$\{\Phi_p(t)\,|\,p \in S_o(k),\ \ \Phi_p(t) \in \mathbb{C}[t]\}$$

satisfying the following conditions:

$$\Phi_p(t) \equiv 1 (\mathrm{mod}\ t^2), \quad \deg \Phi_p \leq d-1, \tag{28}$$

and

$$\ell_p(\chi,t)^{-1} = \Phi_p(t)\ell'_p(\chi,t)^{-1}, \tag{29}$$

where $d = \prod_{j=1}^{r} d_j$, $d_j = \dim \rho_j$, and

$$\ell'_p(\vec{\chi},t) = \det(1-t\rho_1(\sigma_p) \otimes \dots \otimes \rho_r(\sigma_p)). \tag{30}$$

Moreover,

$$L(s,\vec{\chi}) = \prod_{p\in S_o} \ell_p(\vec{\chi},|p|^{-s})^{-1}. \tag{31}$$

<u>Proof</u>. Identity (31) follows from (I.3.23), (27) and (5). Relations (28) and (29) are consequences of (22), (24) and the definitions (27),

(I.3.8).

Let V_j, $1 \leq j \leq r$, be the representation space of ρ_j and let $V = V_1 \otimes \ldots \otimes V_r$, $\rho = \rho_1 \otimes \ldots \otimes \rho_r$, $\chi = \mathrm{tr}\ \rho$. Since

$$V_{jp} = V_j \quad \text{for} \quad p \notin S_o(\rho_j), \quad 1 \leq j \leq r,$$

it follows that

$$S_o(\rho) \subseteq S_o(\vec{\chi}), \tag{32}$$

where

$$S_o(\vec{\chi}) := \bigcup_{j=1}^{r} S_o(\rho_j). \tag{33}$$

Obviously,

$$V_p' = V_{1p} \otimes \ldots \otimes V_{rp}$$

is an invariant subspace of V_p and, since ρ is equivalent to an unitary representation, we have a decomposition

$$V_p = V_p' \oplus V_p'', \quad p \in S_o(k), \tag{34}$$

where V_p'' is the (invariant) subspace of V_p orthogonal to V_p'. Let ρ_p' and ρ_p'' denote the restriction of $\rho(\sigma_p)$ to V_p' and V_p'', respectively, so that

$$\rho(\sigma_p) = \rho_p' \oplus \rho_p'' \quad . \tag{35}$$

It follows from (35) and (I.3.10) that

$$\ell_p(\rho,t) = \ell_p'(\vec{\chi},t)\,\ell_p''(\vec{\chi},t), \tag{36}$$

where

$$\ell_p''(\vec{\chi},t) = \det(1-\rho_p'' t). \tag{37}$$

Finally let

$$L_o(s,\vec{\chi}) = \prod_{p \in S_o(\chi)} \ell_p''(\vec{\chi},|p|^{-s})^{-1}. \tag{38}$$

Theorem 1. The function

$$s \mapsto L(s,\vec{\chi}) \ , \quad s \in \mathbb{C},$$

defined by (26) in the half-plane $\operatorname{Re} s > 1$ can be meromorphically continued to the half-plane $\operatorname{Re} s > \frac{1}{2}$, where it satisfies the following equation:

$$L(s,\vec{\chi}) = L(s,\chi)L(s,\Phi)^{-1}L_o(s,\vec{\chi})^{-1}. \tag{39}$$

Proof. It follows from (31), (29), (21), (36) - (38) if one remarks that

$$\ell''_p(\vec{\chi},t) = 1 \quad \text{for} \quad p \notin S_o(\vec{\chi}) \tag{40}$$

and that, in view of (28), the product (21) converges absolutely for $\operatorname{Re} s > \frac{1}{2}$.

Corollary 2. In notations of Corollary 1,

$$\Phi_p(t) = 1 + \sum_{m=2}^{d-1} t^m b_m(p) \tag{41}$$

with

$$b_m(p) = \sum_{m_1+m_2=m} (-1)^{m_1} \operatorname{tr} \Lambda^{m_1}\Big(\bigotimes_{j=1}^{r} \rho_j(\sigma_p)\Big) \prod_{j=1}^{r} \operatorname{tr}(S^{m_2}\rho_j(\sigma_p)), \tag{42}$$

where m_1 and m_2 range over non-negative integers.

Proof. By (27), (29) and (30),

$$\Phi_p(t) = (\ell_p(\rho_1,t)^{-1} * \ldots * \ell_p(\rho_r,t)^{-1})\det(1-t \bigotimes_{j=1}^{r} \rho_j(\sigma_p)). \tag{43}$$

The statement of this corollary follows from (43) in view of (28) and identities (I.2.16), (I.2.17).

Notations 1. Let Y denote the ring of virtual characters generated by the set of characters

$$X = \{\chi \mid \chi = \operatorname{tr} \rho, \ \ \rho \in R(k)\}.$$

Let us define a sequence of representations

$$\psi_{m_1,m_2} = \Lambda^{m_1}\rho \otimes \left(\bigotimes_{j=1}^{r} S^{m_2}\rho_j\right), \quad m_1 \in \mathbb{N}, \quad m_2 \in \mathbb{N}, \tag{44}$$

and let

$$a_m = \sum_{m_1+m_2=m} (-1)^{m_1} \operatorname{tr} \psi_{m_1,m_2}, \quad m \in \mathbb{N}. \tag{45}$$

Notations 2. Let $a = \sum_{i=1}^{\nu} n_i\chi_i$, $n_i \in \mathbb{Z}$, $\chi_i \in X$. We write then

$$a(p) = \sum_{i=1}^{\nu} n_i\chi_i(p) \quad \text{for} \quad p \in S_o(k),$$

where, as in §I.5 and §I.6,

$$\chi(p) := \operatorname{tr} \rho(\sigma_p) \quad \text{for} \quad p \in S_o(k),\ \chi = \operatorname{tr} \rho,\ \rho \in R(k).$$

Let $H(t) \in Y[t]$, $H(t) = \sum_{j=0}^{\ell} t^j a_j$. Then

$$H_p(t) := \sum_{j=0}^{\ell} t^j a_j(p), \qquad p \in S_o(k). \tag{46}$$

Corollary 3. In notations (44) - (46),

$$\Phi_p(t) = T_p(t) \quad \text{for} \quad p \in S_o(k) \setminus S_o(\vec{\chi}), \tag{47}$$

where

$$T(t) = 1 + \sum_{m=2}^{d-1} t^m a_m. \tag{48}$$

Proof. It follows from (41), (42), (44), (45) and the definitions.

§2. Digression: virtual characters of compact groups.

Let G be a compact group, let μ denote the Haar measure on G normalized by the condition $\mu(G) = 1$ and let R be the set of all the (equivalence classes of) finite dimensional complex (continuous) representations of G. We introduce also the set X_o of all the <u>simple</u> characters of G and the ring Y_o of virtual characters of G, so that

$$Y_o = \{\sum_{\chi} m(\chi)\chi \mid \chi \in X_o,\ m(\chi) \in \mathbb{Z}\},$$

where m ranges over all the functions

$$m: X_o \to \mathbb{Z}$$

with finite support (thus $\{\chi \mid m(\chi) \neq 0\}$ is a finite set for each m).

Given a polynomial

$$P(t) = 1+\sum_{j=1}^{\ell} t^j a_j, \quad a_j \in Y_o, \quad a_\ell \neq 0, \tag{1}$$

let

$$P_g(t) = 1+\sum_{j=1}^{\ell} t^j a_j(g), \qquad g \in G. \tag{2}$$

One defines ℓ functions

$$\alpha_j: G \to \mathbb{C}, \quad 1 \leq j \leq \ell,$$

by the equation

$$P_g(t) = \prod_{j=1}^{\ell} (1-\alpha_j(g)t). \tag{3}$$

Let

$$\gamma = \sup\{|\alpha_j(g)| \mid 1 \leq j \leq \ell,\ g \in G\}. \tag{4}$$

<u>Lemma 1</u>. In notations (1) - (4), the following estimate holds:

$$1 \leq \gamma < \infty. \tag{5}$$

<u>Proof</u>. It follows from (2) - (4) that

$$\gamma^{\ell} \geq \prod_{j=1}^{\ell} |\alpha_j(g)| = |a_{\ell}(g)|. \tag{6}$$

Let

$$a_j = \sum_{\chi \in X_o} m_j(\chi)\chi, \qquad m_j\colon X_o \to \mathbb{Z}, \quad 1 \leq j \leq \ell. \tag{7}$$

It follows from (6) and the orthogonality relations that

$$\gamma^{2\ell} \geq \int_G |a_{\ell}(g)|^2 d\mu(g) = \sum_{\chi \in X_o} |m_{\ell}(\chi)|^2. \tag{8}$$

Relations (1) and (7), (8) show that

$$\gamma \geq 1. \tag{9}$$

Since $|\chi(g)| \leq \chi(1)$ for $g \in G$, we deduce from (7) an estimate:

$$|a_j(g)| \leq B, \quad 1 \leq j \leq \ell, \quad g \in G \tag{10}$$

with

$$B := \max_{1 \leq j \leq \ell} \{ \sum_{\chi \in X_o} |m_j(\chi)|\chi(1)\}.$$

Let $P_g(\alpha^{-1}) = 0$, then (2) gives

$$\alpha^{\ell} + \sum_{j=1}^{\ell} a_j(g)\alpha^{\ell-j} = 0;$$

therefore it follows from (10) that

$$|\alpha| \leq B \sum_{j=0}^{\ell-1} |\alpha|^{-j} = B(1-|\alpha|^{-\ell})(1-|\alpha|^{-1}) \leq B|1-|\alpha|^{-1}|^{-1} \quad ,$$

and therefore $|\alpha| \leq B+1$. Thus $|\alpha_j(g)| \leq B+1$ for $g \in G$, $1 \leq j \leq \ell$, and we get (in view of (4)) an inequality

$$\gamma \leq B+1 \; . \tag{11}$$

Relation (5) follows from (9) and (11).

Definition 1. A polynomial (1) is said to be unitary if $\gamma = 1$.

Corollary 1. Let $P(t) \in Y_o[t]$. The polynomial P is unitary if and only if

$$P_g(\alpha) \neq 0 \quad \text{when} \quad g \in G, \quad \alpha \in \mathbb{C} \quad \text{and} \quad |\alpha| \neq 1. \tag{12}$$

Proof. It follows from (4), (9) and the definition 1.

Notations 1. Let $P(t) \in Y_o[t]$. In notations (1) and (7), let

$$X_1(P) = \{\rho | \rho \in R, \ m_j(\mathrm{tr}\ \rho) \neq 0 \ \text{ for some } \ j\}. \tag{13}$$

Lemma 2. The set $X_1(P)$ is finite for each $P(t)$ in $Y[t]$.

Proof. It follows from the definitions.

We state now the main result of this paragraph.

Theorem 1. Let $P(t) \in Y_o[t]$ and suppose that $P(o) = 1$. There is a sequence of functions

$$\{b_n | b_n : X_1(P) \to \mathbb{Z}, \quad 0 \leq n < \infty\}$$

such that

$$P(t) = \prod_{n=1}^{\infty} \prod_{\varphi \in X_1(P)} \det(1-t^n\varphi)^{b_n(\varphi)} \quad \text{in} \quad Y_o[[t]], \tag{14}$$

and the product

$$P_g(t) = \prod_{n=1}^{\infty} \Big(\prod_{\varphi \in X_1(P)} \det(1-t^n\varphi(g))^{b_n(\varphi)} \Big) \tag{15}$$

converges absolutely for $g \in G$, $|t| < \gamma^{-1}$ with γ defined by (4).

Proof. We construct inductively two sequences: $\{b_n(\varphi)\}$ and

$$\{F_n | F_n(t) \in Y_o[t], \quad 0 \leq n < \infty\}$$

satisfying the following conditions:

$$F_n(t) \equiv P(t) \pmod{t^{n+1}}, \tag{16}$$

and

$$F_n(t) = \prod_{\nu=0}^{n} \prod_{\varphi \in X_1(P)} \det(1-t^{\nu}\varphi)^{b_\nu(\varphi)}. \tag{17}$$

Let $F_o(t) = 1$ and $b_o(\varphi) = 0$, $\varphi \in X_1(P)$; then (16) and (17) are satisfied for $n = 0$ because $P(0) = 1$. Suppose that (16) and (17) hold. Then there is b in Y such that

$$F_n(t) \equiv (1+bt^{n+1})P(t) \tag{18}$$

and, moreover,

$$b = \sum_{\varphi\in X_1(P)} b_{n+1}(\varphi)\operatorname{tr}\varphi, \quad b_{n+1}(\varphi) \in \mathbb{Z}. \tag{19}$$

Relation (19) defines the function b_{n+1}; let

$$F_{n+1}(t) = F_n(t) \prod_{\varphi\in X_1(P)} \det(1-t^{n+1}\varphi)^{b_{n+1}(\varphi)}. \tag{20}$$

One deduces from (16) - (20) the following relations:

$$F_{n+1}(t) \equiv P(t) \pmod{t^{n+2}}, \tag{16'}$$

and

$$F_{n+1}(t) = \prod_{\nu=0}^{n+1} \prod_{\varphi\in X_1(P)} \det(1-t^{\nu}\varphi)^{b_\nu(\varphi)}. \tag{17'}$$

This completes the construction of $\{F_n\}$ and $\{b_n\}$. Identity (14) follows from (16) and (17). Let $\alpha_j\colon G \to \mathbb{C}$, $1 \leq j \leq \ell$, be defined by (3). Then (14) gives:

$$\prod_{j=1}^{\ell} (1-t\alpha_j) = \prod_{n=1}^{\infty} \prod_{\varphi\in X_1(P)} \det(1-t^n\varphi)^{b_n(\varphi)}. \tag{21}$$

We apply to the both sides of (21) the operator

$$-t\frac{\partial}{\partial t} : Y_o[[t]] \to Y_o[[t]]$$

and use a well known identity

$$\log\circ\det = \operatorname{tr}\circ\log. \tag{22}$$

This gives

$$\sum_{j=1}^{\ell} \frac{t\alpha_j}{1-t\alpha_j} = \sum_{n=1}^{\infty} \sum_{\varphi\in X_1(P)} nb_n(\varphi)\operatorname{tr}(t^n\varphi(1-t^n\varphi)^{-1}). \tag{23}$$

Let, for $g \in G$,

$$\sigma(m,g) = \sum_{j=1}^{\ell} \alpha_j(g)^m$$

and

$$h_n(g) = n \sum_{\varphi \in X_1(P)} b_n(\varphi) \operatorname{tr} \varphi(g).$$

It follows from (23) that the identity

$$\sum_{m=1}^{\infty} t^m \sigma(m,g) = \sum_{m,n=1}^{\infty} t^{nm} h_n(g^m), \quad g \in G, \tag{24}$$

holds formally in $C[[t]]$. Relation (24) gives:

$$\sigma(n,g) = \sum_{mm'=n} h_m(g^{m'}), \quad n \in \mathbb{N}, \quad n \neq 0, \tag{25}$$

where m and m' range over positive rational integers. Introducing the Möbius function $\mu: \mathbb{N} \to \{0,-1,1\}$ one obtains from (25):

$$\sum_{r|n} \mu(r)\sigma(\tfrac{n}{r},g^r) = \sum_{r|n} \mu(r) \sum_{mm'=\frac{n}{r}} h_m(g^{m'r}) =$$

$$= \sum_{m|n} h_m(g^{\frac{n}{m}}) \sum_{r|\frac{n}{m}} \mu(r) .$$

Since $\sum_{r|\ell} \mu(r) = 0$ when $\ell > 1$, we get an identity:

$$h_n(g) = \sum_{r|n} \mu(r)\sigma(\tfrac{n}{r},g^r). \tag{26}$$

Let $\tau(n) = \sum_{r|n} 1$ denote the number of positive divisors of n.

Relations (26) and (4) give

$$|h_n(g)| \leq \tau(n)\gamma^{n}\ell ,$$

or

$$\Big| \sum_{\varphi \in X_1(P)} b_n(\varphi) \operatorname{tr} \varphi(g)\Big| \leq \frac{\tau(n)}{n} \gamma^{n}\ell , \quad 1 \leq n < \infty, \quad g \in G. \tag{27}$$

Write, for brevity,

$$A := \sum_{\varphi \in X_1(P)} \log \det(1-t^n\varphi(g))^{b_n(\varphi)}, \quad n \geq 1, \quad g \in G.$$

Applying (22) we get

$$A = -\sum_{m=1}^{\infty} \frac{t^{nm}}{m} \sum_{\varphi\in X_1(P)} b_n(\varphi)\,\mathrm{tr}\,\varphi(g^m) \quad \text{in} \quad \mathbb{C}[[t]]. \tag{28}$$

Let $t \in \mathbb{C}$ and suppose that $|t| < 1$, then (28) and (27) give:

$$|A| \leq \frac{\tau(n)\gamma^{n}\ell}{n} \sum_{m=1}^{\infty} \frac{|t|^{nm}}{m} = \frac{\tau(n)(\gamma|t|)^n}{n}\,\ell \sum_{m=0}^{\infty} \frac{|t|^{nm}}{m+1} . \tag{29}$$

Since $\tau(n) \leq n$ for any n, relation (29) leads to an inequality:

$$|A| \leq \ell(\gamma|t|)^n(1-|t|)^{-1} \quad \text{for} \quad |t| < 1 . \tag{30}$$

Recalling the definition of A we get from (30):

$$\sum_{n\geq M} \Big| \sum_{\varphi\in X_1(P)} \log\det(1-t^n\varphi(g))^{b_n(\varphi)} \Big| \leq \frac{\ell(\gamma|t|)^M}{(1-|t|)(1-\gamma|t|)} \tag{31}$$

for $|t| < \gamma^{-1}$, $t \in \mathbb{C}$. Therefore the product (25) converges absolutely for $|t| < \gamma^{-1}$. This proves theorem 1.

Corollary 2. The sequence $\{b_n(\varphi)\}$ constructed in the course of the proof of theorem 1 satisfies relations (27) and (31).

Corollary 3. If P is unitary, then there is n_o in $\mathbb{N}$ such that

$$b_n(\varphi) = 0 \quad \text{for} \quad n > n_o , \qquad \varphi \in X_1(P), \tag{32}$$

and therefore

$$P(t) = \prod_{n=1}^{n_o} \prod_{\varphi\in X_1(P)} \det(1-t^n\varphi)^{b_n(\varphi)} . \tag{33}$$

Proof. Let

$$a_n(g) = \sum_{\varphi\in X_1(P)} b_n(\varphi)\,\mathrm{tr}\,\varphi(g), \qquad \chi := \mathrm{tr}\,\varphi .$$

The orthogonality relations give:

$$b_n(\varphi) = \int_G a_n(g)\overline{\chi(g)}\,d\mu(g). \tag{34}$$

It follows from (27) and (34) that

$$|b_n(\varphi)| \leq \frac{\tau(n)\ell}{n} \gamma^n \int_G |\chi(g)| d\mu(g). \tag{35}$$

But

$$\left(\int_G |\chi(g)| d\mu(g) \right)^2 \leq \int_G |\chi(g)|^2 d\mu(g) \int_G d\mu(g) = 1. \tag{36}$$

Since P is unitary, relations (35) and (36) give:

$$|b_n(\varphi)| \leq \frac{\tau(n)}{n} \ell, \quad n \in \mathbb{N}, \ n \neq 0, \quad \varphi \in X_1(P). \tag{37}$$

Relation (32) follows from (37) becauce $\frac{\tau(n)}{n} \to 0$ as $n \to \infty$; identity (33) is a consequence of (32).

§3. Analytic continuation of Euler products.

Let $H(t) \in Y[t]$ and suppose that $H(0) = 1$, then the formal Euler product (cf. notations 1.1, notations 1.2 and (21))

$$L(s,H) = \prod_{p \in S_o} H_p(|p|^{-s})^{-1}$$

converges absolutely for $\text{Re } s > 1$. The goal of this paragraph is to prove the following result.

Theorem 1. The function

$$s \mapsto L(s,H) \tag{1}$$

can be analytically continued to a function meromorphic in the half-plane

$$\mathbb{C}_+ = \{s \mid \text{Re } s > 0\}. \tag{2}$$

If H is an unitary polynomial, then (1) can be continued to a function meromorphic in the whole complex plane $\mathbb{C}$.

The following statement is an immediate consequence of theorem 1.

Corollary 1. The scalar product $L(s,\vec{X})$ defined by (1.26) can be meromorphically continued in $\mathbb{C}_+$.

Proof. In view of (1.39) and (1.47), we have

$$L(s,\vec{X}) = L(s,X)L(s,T)^{-1}L_o(s,\vec{X})^{-1} \prod_{p \in S_o(\vec{X})} \Phi_p(|p|^{-s})T_p(|p|^{-s})^{-1}, \tag{3}$$

where $T(t) \in Y[t]$ and $T(0) = 1$. Since $S_o(\vec{X})$ is a finite set, the assertion follows from (3), (1.38), theorem 1 and corollary I.3.2.

Notations 1. Let $\mathfrak{M} \subseteq R(k)$. We write

$$S_o(\mathfrak{M}) = \bigcup_{\rho \in \mathfrak{M}} S_o(\rho). \tag{4}$$

Let $H(t) = 1+\sum_{j=1}^{\ell} a_j t^j$, $a_j = \sum_{i=1}^{q} n_{ji}\chi_i$, $n_{ji} \in \mathbb{Z}$, $\chi_i = \operatorname{tr}\rho_i$, $\rho_i \in R(k)$, $1 \le j \le \ell$, $1 \le i \le q$, and let

$$\mathfrak{M}_H = \{\rho_i | 1 \le i \le q\}. \tag{5}$$

We write then

$$S_o(H) = S_o(\mathfrak{M}_H). \tag{6}$$

It follows from the definitions that $S_o(\mathfrak{M})$ is a finite subset of $S_o(k)$ when $\mathfrak{M}$ is finite; in particular, the set $S_o(H)$ is finite for any H in $Y[t]$. Moreover, there is a finite Galois extension $K|k$ such that

$$\mathfrak{M}_H \subseteq R(K|k). \tag{7}$$

Let $G = W_1(K|k)$, then relation (7) shows that actually

$$H(t) \in Y_o[t], \quad \mathfrak{M}_H = X_1(H), \tag{8}$$

in notations of §2. Furthermore, it follows from (2.2) and (1.46) that

$$H_p(t) = H_\tau(t) \quad \text{for} \quad \tau \in \sigma_p \quad \text{when} \quad p \notin S_o(H). \tag{9}$$

Suppose at first that H is unitary. Then it follows from (2.33) and (9) that there is n_o in $\mathbb{N}$ for which

$$L(s,H) = \prod_{p\in S_o(H)} H_p(|p|^{-s})^{-1} \prod_{p\notin S_o(H)} \prod_{n=1}^{n_o} \prod_{\varphi\in X_1(H)} \det(1-|p|^{-ns}\varphi(\sigma_p))^{-b_n(\varphi)},$$

or recalling (I.3.23) and (I.3.8),

$$L(s,H) = \prod_{p\in S_o(H)} H_p(|p|^{-s})^{-1} \prod_{n=1}^{n_o} \prod_{\varphi\in X_1(H)} L(ns,\operatorname{tr}\varphi)^{b_n(\varphi)} L_1(s,H), \tag{10}$$

where

$$L_1(s,H) = \prod_{p\in S_o(H)} \prod_{n=1}^{n_o} \prod_{\varphi\in X_1(H)} \det(1-|p|^{-ns}\varphi(\sigma_p))^{b_n(\varphi)}. \tag{11}$$

Since the product (11) is finite, identity (10) defines a meromorphic continuation of $L(s,H)$ to the whole complex plane $\mathbb{C}$ (because, by corollary I.3.2, the functions $s \mapsto L(ns, \mathrm{tr}\,\varphi)$ are meromorphic in $\mathbb{C}$). Thus we can assume that H is not unitary. Then it follows from (2.5) that

$$1 < \gamma < \infty \quad . \tag{12}$$

We choose two rational integers N and M satisfying the following conditions:

$$M > 0, \quad N > \gamma^M, \quad N > |p| \quad \text{for each } p \text{ in } S_o(H). \tag{13}$$

Let, for brevity,

$$f_{n,p}(s) = \prod_{\varphi \in X_1(H)} \det(1-|p|^{-ns}\varphi(\sigma_p))^{b_n(\varphi)} \tag{14}$$

with $b_n(\varphi)$ defined by (2.14) when one takes $P(t) = H(t)$. One defines two finite products:

$$Z_N(s) = \prod_{|p|<N} H_p(|p|^{-s})^{-1}, \tag{15}$$

$$R_{N,M}(s) = \prod_{|p|<N}{}' \prod_{n<M} f_{n,p}(s) \ , \tag{16}$$

and two infinite products:

$$U_M(s) = \prod_{n<M} \prod_{p \notin S_o(H)} f_{n,p}(s)^{-1} \ , \tag{17}$$

$$T_{N,M}(s) = \prod_{n \geq M} \prod_{|p| \geq N} f_{n,p}(s)^{-1} \ , \tag{18}$$

where $\prod'_{|p|<N}$ is extended over the finite set

$$\{p \mid p \in S_o(k), \ \ p \notin S_o(H), \ |p| < N\}.$$

By definition,

$$L(s,H) = Z_N(s) \prod_{|p| \geq N} H_p(|p|^{-s})^{-1}.$$

Therefore it follows from (13), (9) and (2.14) that

$$L(s,H) = Z_N(s) \prod_{n=1}^{\infty} \prod_{|p|\geq N} f_{n,p}(s)^{-1} \tag{19}$$

as formal Euler products; relations (16) - (19) imply a formal identity:

$$L(s,H) = Z_N(s)R_{N,M}(s)U_M(s)T_{N,M}(s). \tag{20}$$

Relations (14), (17), (I.3.23) and (I.3.8) give:

$$U_M(s) = \prod_{n<M} \prod_{\varphi\in X_1(H)} L(ns,\mathrm{tr}\ \varphi)^{b_n(\varphi)} \prod_{n<M} \prod_{p\in S_o(H)} f_{n,p}(s). \tag{21}$$

Lemma 1. The function

$$s \mapsto Z_N(s)R_{N,M}(s)U_M(s)$$

is meromorphic in $\mathbb{C}$.

Proof. It follows from (14) - (16), (21) and corollary I.3.2.

Lemma 2. The product $T_{N,M}(s)$ defined by (18) converges absolutely when $\mathrm{Re}\ s > \frac{1}{M}$, $s \in \mathbb{C}$.

Proof. Let $M > 0$, $N > \gamma^M$ and suppose that $|p| \geq N$. Then

$$\gamma|p|^{-n\ \mathrm{Re}\ s} < 1 \quad \text{when} \quad \mathrm{Re}\ s > \frac{1}{M},\ n \geq 1.$$

Therefore it follows from (13), (14) and (2.31) with $t = |p|^{-s}$, $g \in \sigma_p$ that

$$\sum_{n\geq M} |\log f_{n,p}(s)| \leq A|p|^{-M\ \mathrm{Re}\ s} \text{ for } \mathrm{Re}\ s > \frac{1}{M},\ |p| \geq N, \tag{22}$$

where

$$A:= \ell\gamma^M(1-N^{-1/M})^{-1}(1-\gamma N^{-1/M})^{-1} \quad .$$

Since

$$\sum_{|p|\geq N} |p|^{-M\ \mathrm{Re}\ s} \leq \zeta_k(M(\mathrm{Re}\ s)) \quad \text{for} \quad \mathrm{Re}\ s > \frac{1}{M}$$

estimate (22) proves the lemma.

Let

$$\mathbb{C}_{1/M} = \{s \mid \mathrm{Re}\ s > \frac{1}{M}, \quad s \in \mathbb{C}\}. \tag{23}$$

In view of Lemma 1 and Lemma 2, identity (20) defines a meromorphic continuation of (1) to $\mathbb{C}_{1/M}$. Since

$$\mathbb{C}_+ = \bigcup_{M=1}^{\infty} \mathbb{C}_{1/M} , \tag{24}$$

the proof of theorem 1 is completed.

§4. The natural boundary of L(s,H).

Making use of a prime number theorem to be proved in §6 we prove here the following statement.

Theorem 1. If H is not unitary, the function (3.1) has a natural boundary $\mathbb{C}_o = \{s | \mathrm{Re}\ s = 0,\ s \in \mathbb{C}\}$ and cannot be continued analytically to $\mathbb{C}_- = \{s | \mathrm{Re}\ s < 0,\ s \in \mathbb{C}\}$.

Notations 1. Let $\mathfrak{M} \subseteq R(k)$ and let

$$\check{\mathfrak{M}} = \{\chi | \chi = \mathrm{tr}\ \rho,\ \rho \in \mathfrak{M}\} \tag{1}$$

Let, in notations of (3.4),

$$\mathcal{P}_{\mathfrak{M}}(g,\varepsilon) = \{p | p \notin S_o(\mathfrak{M}),\ |\chi(\sigma_p)-\chi(g)| < \varepsilon \ \text{ for each } \chi \text{ in } \check{\mathfrak{M}}\} \tag{2}$$

and let

$$\mathcal{P}_{\mathfrak{M}}(g,\varepsilon;x_1,x_2) = \{p | p \in \mathcal{P}_{\mathfrak{M}}(g,\varepsilon),\ x_1 \leq |p| < x_2\}, \tag{3}$$

where $g \in W(k)$, $\varepsilon > 0$, $x_2 > x_1 \geq 2$. Write

$$P_{\mathfrak{M}}(g,\varepsilon;x_1,x_2) := \mathrm{card}\ \mathcal{P}_{\mathfrak{M}}(g,\varepsilon;x_1,x_2). \tag{4}$$

Theorem 2. Let $\mathfrak{M}$ be a finite subset of $R(k)$ and let $0 < \varepsilon < 1$. There are c_j, $1 \leq j \leq 4$, satisfying the following conditions:

$c_j > 0$ for $j \leq 3$, $c_4 \in \mathbb{R}$, and

$$P_{\mathfrak{M}}(g,\varepsilon;x_1,x_2) \geq c_1\varepsilon^{c_2} \int_{x_1}^{x_2} \frac{du}{\log u} + c_4 x_2 \exp(-c_3\sqrt{\log x_2}), \tag{5}$$

where c_j, $1 \leq j \leq 4$, does not depend on ε, g, x_1, x_2 (but may depend on $\mathfrak{M}$).

Relation (5) is proved in §6; we use it here in the proof of theorem 1 which follows. Let us retain notations and assumptions (3.12) - (3.18) assuming, in particular, that H is not unitary.

Lemma 1. The following relations are satisfied:

$$R_{N,M}(s) \neq 0 \quad \text{for} \quad s \notin \mathbb{C}_0 , \tag{6}$$

and

$$T_{N,M}(s) \neq 0 \quad \text{for} \quad s \in \mathbb{C}_{1/M} . \tag{7}$$

Proof. Since any representation in $R(k)$ is equivalent to an unitary one, it follows from (3.14) that

$$f_{n,p}(s) \neq 0 \quad \text{when} \quad s \notin \mathbb{C}_0 . \tag{8}$$

Relation (6) follows from (8) and (3.16). Relation (7) follows from lemma 3.2.

Let $t_o \in \mathbb{R}$, $\delta > 0$, $\nu > 0$, and let

$$D_\nu(\delta, t_o) = \{s \mid s \in \mathbb{C}, \frac{1}{\nu+1} < \operatorname{Re} s \leq \frac{1}{\nu} , t_o < \operatorname{Im} s \leq t_o + \delta\}. \tag{9}$$

Let

$$a_1(\nu; \delta, t_o) = \operatorname{card}\{s \mid U_M(s) = 0, s \in D_\nu(\delta, t_o)\}, \tag{10}$$

and let

$$a_2(\nu; \delta, t_o) = \operatorname{card}\{s \mid Z_N(s)^{-1} = 0, \quad s \in D_\nu(\delta, t_o)\} . \tag{11}$$

Lemma 2. There is $A_H(\delta, t_o)$ such that if $\nu \geq 2$ and $M > 0$, then

$$a_1(\nu; \delta, t_o) \leq A_H(\delta, t_o)\nu^2 \log \nu. \tag{12}$$

Proof. It follows from (3.21), (8) and (10) that

$$a_1(\nu; \delta, t_o) \leq \sum_{n<M} \sum_{\varphi \in X_1(H)} \operatorname{card}\{s \mid s \in D_\nu(\delta, t_o), L(ns, \operatorname{tr} \varphi) = 0\} . \tag{13}$$

For $\varphi \in R(k)$, let

$$c(n,\varphi) = \mathrm{card}\{s \mid \frac{n}{\nu+1} < \mathrm{Re}\, s \leq \frac{n}{\nu},\ nt_o < \mathrm{Im}\, s \leq n(t_o+\delta); L(s,\mathrm{tr}\ \varphi) = 0\},$$

and let

$$N(\varphi,T) = \mathrm{card}\{s \mid 0 \leq \mathrm{Re}\, s \leq 1,\ |\mathrm{Im}\, s| \leq T,\ L(s,\mathrm{tr}\ \varphi) = 0\}.$$

Since $L(s,\mathrm{tr}\ \varphi) \neq 0$ for $\mathrm{Re}\, s > 1$, $\varphi \in R(k)$, it follows from (9) that (13) may be rewritten as follows:

$$a_1(\nu;\delta,t_o) \leq \sum_{n<\nu+1} \sum_{\varphi\in X_1(H)} c(n,\varphi) . \tag{14}$$

Obviously,

$$c(n,\varphi) \leq N(\varphi,n(t_o+\delta)) - N(\varphi,nt_o). \tag{15}$$

Relations (15), (I.6.19), and (I.3.26) show that

$$c(n,\varphi) \leq (n\delta+1)\ \log(n|t_o|+n\delta+3)c_1(\varphi), \tag{16}$$

where $c_1(\varphi)$ depends solely on φ. Since the set $X_1(H)$ is finite, relation (12) follows from (14) and (16).

Lemma 3. There are two positive real numbers ε_o and β such that if $0 < \varepsilon < \varepsilon_o$ and $p \in \mathcal{P}_{X_1(H)}(g,\varepsilon^\beta)$, then the polynomial $H_p(t)$ has a root $\kappa(p)^{-1}$ for which

$$|\log|\kappa(p)| - \log\gamma| < \varepsilon. \tag{17}$$

Proof. Since $G := W_1(K|k)$ is a compact group, it follows from (3.8), (2.3) and (2.4) (with $P = H$) that

$$H_g(t) = (1-\alpha(g)t)^{\beta_1} H_g^{(1)}(t),\quad \beta_1 \geq 1,\quad |\alpha(g)| = \gamma, \tag{18}$$

and

$$H_g^{(1)}(\alpha(g)^{-1}) \neq 0 \tag{19}$$

for an appropriate element g in G. Let $0 < \varepsilon_1 < 1$ and suppose that

$$H_g^{(1)}(t) \neq 0 \quad \text{when} \quad |t-\alpha(g)^{-1}| \leq \varepsilon_1 \tag{20}$$

(relation (19) shows that there is a real number ε_1 satisfying these conditions). By (20),

$$|H_g^{(1)}(t)| \geq w \quad \text{when} \quad |t-\alpha(g)^{-1}| \leq \varepsilon_1 \tag{21}$$

for some w in $\mathbb{R}_+$. It follows from (18) and (21) that if $0 < \varepsilon \leq \varepsilon_1$, then

$$|H_g(t)| \geq w\gamma^{\beta_1}\varepsilon^{\beta_1} \quad \text{on the circle} \quad |t-\alpha(g)^{-1}| = \varepsilon. \tag{22}$$

Let $p \in \mathcal{P}_{X_1(H)}(g,\varepsilon)$, then relations (3.5), (3.8) and (2) imply that

$$|a_j(g)-a_j(\sigma_p)| < w_1\varepsilon \; , \quad 1 \leq j \leq \ell \tag{23}$$

for an appropriate w_1 in $\mathbb{R}_+$. Let

$$H_p(t) = H_g(t) + h_p(t). \tag{24}$$

It follows from (23) and (24) that

$$|h_p(t)| \leq w_1\varepsilon \sum_{j=1}^{\ell} |t|^j \; ,$$

and therefore, by (18),

$$|h_p(t)| \leq \ell w_1(1+\gamma^{-1})^{\ell}\varepsilon \quad \text{when} \quad |t-\alpha(g)^{-1}| \leq 1. \tag{25}$$

In particular, if $0 < \varepsilon \leq 1$ and $p \in \mathcal{P}_{X_1(H)}(g,\varepsilon^{\beta_1+1})$, then

$$|h_p(t)| \leq \ell w_1(1+\gamma^{-1})^{\ell}\varepsilon^{\beta_1+1} \quad \text{on the circle} \quad |t-\alpha(g)^{-1}| = \varepsilon. \tag{26}$$

Relations (22) and (26) show that one can find ε_2 in $\mathbb{R}_+$ satisfying the following condition:

$$|h_p(t)| < H_g(t) \quad \text{on the circle} \quad |t-\alpha(g)^{-1}| = \varepsilon \tag{27}$$

for each p in $\mathcal{P}_{X_1(H)}(g,\varepsilon^{\beta_1+1})$ and each ε in the interval $0<\varepsilon<\varepsilon_2$. By a well known lemma (cf. e.g., [88], §3.42), it follows from (24) and (27) that $H_p(t)$ has a root $\kappa(p)^{-1}$ in the disk $|t-\alpha(g)^{-1}|\leq\varepsilon$. Thus there is a positive real number ε_3 satisfying the following condition: if $p\in\mathcal{P}_{X_1(H)}(g,\varepsilon^{\beta_1+1})$ and $0<\varepsilon\leq\varepsilon_3$, then

$$H_p(\kappa(p)^{-1})=0 \quad\text{and}\quad |\log|\kappa(p)|-\log\gamma|<2\gamma\varepsilon \tag{28}$$

for some $\kappa(p)$ in $\mathbb{C}$. The statement of the lemma (with $\beta=\beta_1+2$) is a consequence of (28).

<u>Lemma 4</u>. There are two real numbers $A_H^{(1)}(\delta,t_o)$ and $A_H^{(2)}(\delta,t_o)$ satisfying the following conditions: $A_H^{(1)}(\delta,t_o)>0$, and if $N>\gamma^{\nu+1}$ and $\nu>A_H^{(2)}(\delta,t_o)$, then

$$a_2(\nu;\delta,t_o)>A_H^{(1)}(\delta,t_o)\nu^3.$$

<u>Proof</u>. Let $0<\varepsilon<\varepsilon_o$; for each p in $_{X_1(H)}(g,\varepsilon^{\beta})$ we choose, by lemma 3, a complex number $\kappa(p)$ satisfying (17) and such that

$$H_p(\kappa(p)^{-1})=0.$$

Let us assume that

$$\nu>\frac{2\pi}{\delta\log\gamma}$$

and let

$$Q(\nu)=\mathcal{P}_{X_1(H)}(g,\varepsilon^{\beta};(\gamma\exp\varepsilon)^{\nu},(\gamma\exp(-\varepsilon))^{\nu+1}).$$

Then

$$\frac{2\pi}{\log|p|}<\delta \quad\text{for}\quad p\in Q(\nu). \tag{*}$$

Let $p\in Q(\nu)$ and let $\kappa(p)=|p|^{s(p)}$, $s(p)\in\mathbb{C}$. It follows from (17) that

$$\frac{1}{\nu+1}<\operatorname{Re}s(p)<\frac{1}{\nu},$$

and since Im $s(p)$ is defined modulo $\frac{2\pi}{\log|p|}\mathbb{Z}$ only we may assume, in view of (*), that

$$t_o < \text{Im } s(p) \leq t_o+\delta.$$

Thus $s(p) \in D_\nu(\delta,t_o)$, and we obtain a relation (cf. (3.15)):

$$Z_N(s(p))^{-1} = 0, \; s(p) \in D_\nu(\delta,t_o) \quad \text{for} \quad p \in Q(\nu), \tag{29}$$

as soon as

$$N > \gamma^{\nu+1}. \tag{30}$$

Choose

$$\varepsilon = \nu^{-4}, \quad \lambda = 2\varepsilon(\nu+1), \tag{31}$$

and let

$$Q_j(\nu) = \mathcal{P}_{X_1(H)}(g,\varepsilon^\beta;\gamma^\nu \exp(j\lambda+\nu\varepsilon),\gamma^\nu \exp((j+1)\lambda+\nu\varepsilon)). \tag{32}$$

Let further

$$J = \{j|j \in \mathbb{N}, \; \gamma^\nu \exp((j+1)\lambda+\nu\varepsilon) < (\gamma \exp(-\varepsilon))^{\nu+1}\}. \tag{33}$$

It follows from (32) and (33) that

$$Q_j(\nu) \subseteq Q(\nu) \quad \text{for} \quad j \in J. \tag{34}$$

Clearly if $p \in Q_j(\nu)$ and $q \in Q_{j'}(\nu)$ then

$$|\log|p| - \log|q|| > 2\varepsilon(\nu+1)(|j-j'|-1),$$

in view of (31) and (32). On the other hand, it follows from (17) that if $p \in Q(\nu)$, $q \in Q(\nu)$ and $s(p) = s(q)$ then

$$|\log|p| - \log|q|| < \frac{2\varepsilon}{\text{Re } s(p)} < 2\varepsilon(\nu+1).$$

Thus, by (34),

$$s(p) \neq s(q) \quad \text{whenever} \quad p \in Q_j(\nu), \; q \in Q_{j'}(\nu), \; |j-j'| \geq 2, \; j \in J, \; j' \in J. \tag{35}$$

It follows from (5) and (32) that

$$|Q_j(\nu)| \geq c_1 \varepsilon^{c_2\beta} \int_{x_1}^{x_2} \frac{du}{\log u} + c_4 x_2 \exp(-c_3\sqrt{\log x_2}), \tag{36}$$

where

$$x_1 = \gamma^\nu \exp(j\lambda+\nu\varepsilon), \; x_2 = \gamma^\nu \exp(\lambda(j+1)+\nu\varepsilon). \tag{37}$$

Let $j \in J$. By (31) and (37),

$$x_2 - x_1 \geq \lambda\gamma^\nu \quad ;$$

moreover, it follows from (33) and (37) that

$$\nu \log \gamma \leq \log x_2 \leq (\nu+1)\log \gamma.$$

Therefore (36) gives:

$$|Q_j(\nu)| \geq c_1 \varepsilon^{c_2\beta} (\lambda\gamma^\nu)(\nu+1)^{-1}(\log \gamma)^{-1} - |c_4|\gamma^{\nu+1}\exp(-c_3\sqrt{\nu\log \gamma}). \tag{38}$$

Let

$$c_5 = 2c_1(\log \gamma)^{-1}, \; c_6 = 4(c_2\beta+1), \; c_7 = c_3\sqrt{\log \gamma}, \; c_8 = \gamma|c_4|.$$

In these notations, relations (38) and (31) give:

$$|Q_j(\nu)| \geq \gamma^\nu (c_5\nu^{-c_6} - c_8 \exp(-c_7\sqrt{\nu})),$$

or taking into account that $c_5 > 0$ (cf. theorem 2 and (3.12))

$$|Q_j(\nu)| \geq c_5 \nu^{-c_6}\gamma^\nu (1 - \frac{c_8}{c_5} \nu^{c_6} \exp(-c_7\sqrt{\nu})). \tag{39}$$

Relation (39) shows that there is $c_9(H)$ which does not depend on ν and which satisfies the following relation:

$$|Q_j(\nu)| \geq 1 \quad \text{for} \quad \nu > c_9(H), \; j \in J. \tag{40}$$

It follows from (11), (35), (29), (34) and (40) that

$$a_2(\nu;\delta,t_o) \geq \frac{1}{2}|J|, \tag{41}$$

as soon as

$$\nu > 2\pi(\delta \log \gamma)^{-1}, \ \nu > c_9(H), \ 0 < \varepsilon < \varepsilon_o, \ N > \gamma^{\nu+1}. \tag{42}$$

The assertion of lemma 4 is an immediate consequence of the relations (31), (33), (41) and (42).

Proposition 1. The line $\mathbb{C}_o = \{s \,|\, \mathrm{Re}\ s = 0\}$ is contained in the closure of the set $\{s \,|\, s \in \mathbb{C}_+,\ L(s,H)^{-1} = 0\}$ of poles of the function (3.1) in $\mathbb{C}_+$.

Proof. Let $it_o \in \mathbb{C}_o$, i.e. $t_o \in \mathbb{R}$, and $U(t_o)$ be an open subset in $\mathbb{C}$ containing the point it_o. Choose $\delta > 0$ and $\nu_o > 0$ in such a way that

$$D_\nu(\delta,t_o) \subseteq U(t_o) \quad \text{for} \quad \nu > \nu_o. \tag{43}$$

It follows from lemma 2 and lemma 4 that there is $A_H^{(3)}(\delta,t_o)$ satisfying the relation:

$$a_2(\nu;\delta,t_o) > a_1(\nu;\delta,t_o) \text{ whenever } N > \gamma^{\nu+1},\ M > 0,\ \nu > A_H^{(3)}(\delta,t_o). \tag{44}$$

Let

$$\nu > \max\{\nu_o, A_H^{(3)}(\delta,t_o)\},\ M = \nu+1, \tag{45}$$

and suppose that (3.13) holds. Relations (3.20), (43) - (45), and (9) - (11) combined with lemma 1 show that there is a complex number s satisfying the conditions:

$$s \in U(t_o),\ L(s,H)^{-1} = 0,$$

and the assertion of proposition 1 follows.

Theorem 1 is an immediate consequence of proposition 1.

Remark 1. In the proof of theorem 1 we have made essential use of theorem 2 (cf. proof of lemma 4).

§5. Explicit calculations related to scalar products.

Returning to notations of §1, let k_j be a finite extension of k, let $d_j = [k_j:k]$ and let $\psi_j \in gr(k_j)$, $1 \leq j \leq r$. Let

$$L(s,\psi_j) = \sum_{\mathfrak{m}\in I_o(k)} a(\mathfrak{m},\psi_j)|\mathfrak{m}|^{-s}, \quad \text{Re } s > 1, \tag{1}$$

where

$$a(\mathfrak{m},\psi_j) = \sum_{\mathfrak{a}\in I_o(k_j)}{}' \psi_j(\mathfrak{a}), \quad N_{k_j/k}\,\mathfrak{a} = \mathfrak{m}, \tag{2}$$

is a finite sum extended over integral ideals $\mathfrak{a}$ in $I_o(k_j)$ subject to the condition $N_{k_j/k}\,\mathfrak{a} = \mathfrak{m}$. We define the scalar product $L(s,\vec{\psi})$ of L-functions (1) by an identity

$$L(s,\vec{\psi}) = \sum_{\mathfrak{m}\in I_o(k)} |\mathfrak{m}|^{-s} \prod_{j=1}^{r} a(\mathfrak{m},\psi_j), \quad \text{Re } s > 1, \tag{3}$$

where $\vec{\psi} := (\psi_1,\ldots,\psi_r)$.

Proposition 1. Let $\rho_j = \mathrm{Ind}_{W(k_j)}^{W(k)}\psi_j$, $1 \leq j \leq r$. Then $d_j = \dim \rho_j$, $1 \leq j \leq r$, and, in notations of (1.26),

$$L(s,\vec{\psi}) = L(s,\vec{\chi}),\ \vec{\chi} := (\chi_1,\ldots,\chi_r),\ \chi_j = \mathrm{tr}\,\rho_j. \tag{4}$$

Proof. It follows from (1.26), (1.25), (1) - (3) and (I.3.25).

We introduce two conditions:

$$d_1 \geq d_2 \geq \ldots \geq d_r \geq 1 \tag{5}$$

and

$$(r \geq 3 \wedge d_3 \geq 2) \vee (r \geq 2 \wedge d_1 \geq 3 \wedge d_2 \geq 2). \tag{6}$$

As in §1, let

$$\prod_{j=1}^{r} d_j = d. \tag{7}$$

The following two lemmas on Hadamard convolution are valid in the generality of §1. Let F be a (commutative) field of characteristic zero.

Lemma 1. Let $a_j \in F$, $b_j \in F$, $j = 1,2$. Then

$$\frac{1}{(1-a_1t)(1-a_2t)} * \frac{1}{(1-b_1t)(1-b_2t)} = (1-a_1a_2b_1b_2t^2) \prod_{1\leq i,j\leq 2} (1-a_ib_jt)^{-1}. \quad (8)$$

Proof. Suppose that either $a_1 \neq a_2$ or $b_1 \neq b_2$, say, $a_1 \neq a_2$. Then

$$\frac{1}{(1-a_1t)(1-a_2t)} = \frac{1}{a_1-a_2}\left(\frac{a_1}{1-a_1t} - \frac{a_2}{1-a_2t}\right),$$

and therefore the expression in the left hand side of (8) is equal to

$$\frac{1}{a_1-a_2}\left[\frac{a_1}{(1-a_1b_1t)(1-a_1b_2t)} - \frac{a_2}{(1-a_2b_1t)(1-a_2b_2t)}\right],$$

and we get (8). If $a_1 = a_2 = a$, $b_1 = b_2 = b$, we get from (1.6):

$$\frac{1}{(1-at)^2} * \frac{1}{(1-bt)^2} = D(t(1-abt)^{-2}) = \frac{1-(abt)^2}{(1-abt)^4}.$$

This proves the lemma.

Lemma 2. Let $f_j(t) = (1-t)^{-d_j}$, $d_j \geq 1$, $1 \leq j \leq r$. Then

$$(f_1*\ldots*f_r)(t) = (1-t)^{-m}h(t), \quad (9)$$

where

$$h(t) \in F[t], \quad m = \sum_{j=1}^{r} d_j+1-r. \quad (10)$$

Moreover,

$$\deg h \leq m-d_1 \quad (11)$$

and

$$h(t) \equiv 1+(d-m)t \pmod{t^2}. \quad (12)$$

Proof. Let $m_j = d_j-1$, $1 \leq j \leq r$. By lemma 1.1,

$$(f_1*\ldots*f_r)(t) = (\prod_{j=2}^{r} m_j!)^{-1} D^{m_r} t^{m_r} \ldots D^{m_2} t^{m_2} (1-t)^{-d_1}. \tag{13}$$

Relations (9) and (10) are direct consequences of (13). Inequality (11) follows from (1.10) applied r-1 times to (13). By lemma 1.3 and lemma 1.4,

$$(f_1*\ldots*f_r)(t)(1-t)^d \equiv 1 (\mathrm{mod}\ t^2). \tag{14}$$

Relation (12) follows from (14) and (9). This proves the lemma.

Lemma 3. In notations of lemma 2, let $F = \mathbb{C}$ and suppose that conditions (5) and (6) are satisfied. Then there is β in $\mathbb{C}$ such that

$$h(\beta) = 0 \text{ and } |\beta| < 1. \tag{15}$$

Proof. In view of (11) and (12), one can write

$$h(t) = \prod_{j=1}^{m-d_1} (1+\beta_j t), \quad \sum_{j=1}^{m-d_1} \beta_j = d-m, \ \beta_j \in \mathbb{C}. \tag{16}$$

It follows from (16) that

$$\max_{1 \le j \le m-d_1} |\beta_j| \ge \frac{d-m}{m-d_1} \quad . \tag{17}$$

On the other hand,

$$d = \prod_{j=1}^{r} (1+m_j) \ge m + \sum_{1 \le i < j \le r} m_i m_j. \tag{18}$$

If $r \ge 2$, $d_1 \ge 3$, $d_2 \ge 2$, then (18) gives:

$$d-m \ge m_1 \sum_{j=2}^{r} m_j \ge 2(m-d_1) > 0 ;$$

if $r \ge 3$ and $d_1 \ge d_2 \ge d_3 \ge 2$, then

$$d-m \ge \frac{r(r-1)}{2} \ge r,$$

so that relation $d_1 = 2$ combined with (5) gives:

$$d-m > m-d_1 > 0. \tag{19}$$

Thus (5) and (6) imply (19). The assertion of the lemma follows from (17) and (19).

Now we turn to a detailed investigation of the properties of the scalar products (1.26) and (3). Let us assume, without loss of generality, that (5) holds.

Theorem 1. The functions

$$s \mapsto L(s,\vec{\chi}) \quad \text{and} \quad s \mapsto L(s,\vec{\psi}) \tag{20}$$

defined by (3) and (1.26), respectively, can be meromorphically continued to $\mathbb{C}_+$. If (6) holds, then $\mathbb{C}_o$ is the natural boundary of the functions (20) and consequently these functions allow for no analytic continuation to $\mathbb{C}_-$. If (6) doesn't hold, the functions (20) can be meromorphically continued to the whole complex plane $\mathbb{C}$.

Proof. The meromorphic continuation to $\mathbb{C}_+$ has been proved for $L(s,\vec{\chi})$ in §3 (see Corollary 3.1); this continuation for $L(s,\vec{\psi})$ can be obtained from Corollary 3.1 and Proposition 1. Suppose that (6) holds. By construction (cf. (1.43), (1.47) and (1.48)), we have

$$T_g(t) = (\ell_g(\rho_1,t)^{-1} * \ldots * \ell_g(\rho_r,t)^{-1})\,\ell_g(\rho,t), \tag{21}$$

where

$$\ell_g(\varphi,t) := \det(1-t\varphi(g)), \quad \varphi \in R(k), \quad g \in W(k). \tag{22}$$

In particular,

$$T_1(t) = ((1-t)^{-d_1} * \ldots * (1-t)^{-d_r})(1-t)^d,$$

and it follows from (19), (9) and (15) that $T(t)$ is not unitary. Equations (3.3), (4) and theorem 4.1 show that $\mathbb{C}_o$ is the natural boundary for the functions (20). If (6) is not valid, then either $(r=1 \vee d_1=1)$

in which case $L(s,\vec{\chi})$ coincides with an Artin-Weil L-function, or

$$(r = d_1 = d_2 = 2) \vee (r \geq 3 \wedge d_1 = d_2 = 2 \wedge d_3 = 1). \tag{23}$$

In the latter case it follows from (21) - (23) and lemma 1 that

$$T_g(t) = 1-t\ \det(\rho(g)),\quad g \in W(k);$$

therefore

$$L(s,T) = L(s,\lambda),\quad \lambda := \det\ \rho,\quad \lambda \in gr(k).$$

Thus if (6) is not valid, the functions (20) are meromorphic in $\mathbb{C}$, by (3.3) and (4).

In the rest of this paragraph we are concerned with the scalar product of Hecke L-functions defined by (3). To simplify our notations, let us assume that the degrees of the fields k_j, $1 \leq j \leq r$, satisfy (5). If (6) does not hold, the function $L(s,\vec{\psi})$ can be explicitly evaluated. To be more precise we prove the following statement.

<u>Proposition 2</u>. If $r = 1$ or $k_j = k$ for $1 < j \leq r$, then

$$L(s,\vec{\psi}) = L(s,\psi_1 \prod_{j>1} \psi_j \circ N_{k_1/k}). \tag{24}$$

If $k_1 = k_2$ and $d_1 = d_2 = 2$, then assuming that either $r = 2$ or $d_3 = 1$ we have the following identity:

$$L(s,\vec{\psi}) = L(s,\chi)L(s,\chi')L(2s,\chi_o)^{-1}L_o(s,\chi), \tag{25}$$

where

$\chi_o \in gr(k)$, $\chi_o(p) = \psi_1(p)\psi_2'(p)$ for $p \in S_o(k)$, $\psi_2' = \psi_2 \prod_{j>2} \psi_j \circ N_{K/k}$, $\chi = \psi_1\psi_2'$,

$\chi' = \psi_1\psi_2'^{-1}(\psi_2' \circ N_{K/k})$, $K := k_i\ (i = 1,2)$, so that $\chi \in gr(K)$, $\chi' \in gr(K)$;

$L_o(s,\chi) = \prod_{p|D(K|k)} (1+\chi(\mathfrak{p})|p|^{-s})^{-1}$, $\mathfrak{p} \in S_o(K)$, $\mathfrak{p}^2 = p$, $p \in S_o(k)$.

If $d_1 = d_2 = 2$, $(r = 2 \vee d_3 = 2)$, and $k_1 \neq k_2$, then

$$L(s,\vec{\psi}) = L(s,\chi)L(2s,\chi_o)^{-1}L_o(s,\vec{\psi}), \tag{26}$$

where $\chi = (\psi_1 \cdot N_{K/k_1})(\psi_2 \cdot N_{K/k_2}) \prod_{j>2} \psi_j \cdot N_{K/k}$, $K = k_1 \cdot k_2$ is the composite of k_1 and k_2, $\chi_o \in gr(k)$, $\chi_o(p) = \psi_1(p)\psi_2(p)\kappa(p) \prod_{j>2} \psi_j(p^2)$ for $p \in S_o(k)$, κ denotes the character belonging to the quadratic extension $k_3|k$, $k_3 \neq k_i$ for $i = 1,2,$ so that

$$\kappa(p) = \begin{cases} 1 & \text{if } p \text{ splits in } k_3 \\ -1 & \text{if } p \text{ remains prime in } k_3 \\ 0 & \text{if } p \text{ is ramified in } k_3|k \end{cases},$$

and

$$L_o(s,\vec{\psi}) = \prod_{p|D} (1-\chi_o(p)|p|^{-2s})(1-\kappa(p)\psi_1(\mathfrak{p}_1)\psi_2(\mathfrak{p}_2)|p|^{-s}),$$

$$D = (D(k_1|k),\ D(k_2|k)),\ p = \mathfrak{p}_1^2 = \mathfrak{p}_2^2,\ \mathfrak{p}_i \in S_o(k_i),\ i = 1,2,$$

$p \in S_o(k)$. Here $D(E|F)$ denotes the relative discriminant of a field extension $E|F$.

<u>Proof</u>. Identity (24) is an immediate consequence of (3). Let $d_1 = d_2 = 2$. If $r > 2$ but $k_j = k$ for $j \geq 3$, we can define a character

$$\psi_2' = \psi_2 \prod_{j \geq 3} \psi_j \cdot N_{k_2/k},\quad \psi_2' \in gr(k_2).$$

It follows from (3) that

$$L(s,\vec{\psi}) \doteq L(s,(\psi_1,\psi_2')).$$

Therefore it suffices to consider the case $d_1 = d_2 = r = 2$. In this case (1.5) gives:

$$L(s,\vec{\psi}) = \prod_{p \in S_o(k)} (\ell_{1p} * \ell_{2p})(|p|^{-s}), \tag{27}$$

where

$$\ell_{ip}(t) = \prod_{\substack{\mathfrak{p} \in S_o(k_i) \\ \mathfrak{p}|p}} (1-\psi_i(\mathfrak{p})t^{f_i(p)})^{-1},\quad N_{k_i/k}\mathfrak{p} = p^{f_i(p)},\ i=1,2. \tag{28}$$

If $k_1 = k_2 =: K$, it follows from (28) that

$$\ell_{1p}(t)*\ell_{2p}(t) = (1-(\psi_1\psi_2)(\mathfrak{p})t)^{-1} \quad \text{when} \quad p = \mathfrak{p}^2, \quad \mathfrak{p} \in S_o(K). \tag{29}$$

Making use of (8) we get from (28) two other identities:

$$\ell_{1p}(t)*\ell_{2p}(t) = (1-(\psi_1\psi_2)(p)t^2)^{-1} \quad \text{when} \quad p \text{ remains prime in } K, \tag{30}$$

and

$$\ell_{1p}(t)*\ell_{2p}(t) = (1-(\psi_1\psi_2)(p)t^2) \prod_{1\le i,j\le 2} (1-\psi_1(\mathfrak{p}_i)\psi_2(\mathfrak{p}_j)t)^{-1} \tag{31}$$

when p splits in K, say, $p = \mathfrak{p}_1\mathfrak{p}_2$, $\mathfrak{p}_1 \ne \mathfrak{p}_2$, $\mathfrak{p}_i \in S_o(K)$. Identity (25) follows from (27) - (31). Suppose now that $k_1 \ne k_2$, $[k_j:k] = 2$ for $j = 1,2$, and let $K = k_1 \cdot k_2$, $[k_3:k] = 2$, $k_3 \subset K$, $k_3 \ne k_1$, $k_3 \ne k_2$. It follows from (28) that

$$\ell_{1p}(t)*\ell_{2p}(t) = (1-\psi_1(\mathfrak{p})\psi_2(\mathfrak{q})t)^{-1} \quad \text{when} \quad p = \mathfrak{p}^2 = \mathfrak{q}^2, \ \mathfrak{p} \in S_o(k_1),$$

$$\mathfrak{q} \in S_o(k_2). \tag{32}$$

On the other hand, if $p|D$ then

$$p = (\mathfrak{P}_1\mathfrak{P}_2)^2, \ \mathfrak{P}_1 \ne \mathfrak{P}_2, \ \mathfrak{P}_j \in S_o(K) \quad \text{when} \quad \kappa(p) = 1, \tag{33.1}$$

$$p = \mathfrak{P}^2, \ \mathfrak{P} \in S_o(K) \quad \text{when} \quad \kappa(p) = -1, \tag{33.2}$$

$$p = \mathfrak{P}^4, \ \mathfrak{P} \in S_o(K) \quad \text{when} \quad \kappa(p) = 0. \tag{33.3}$$

Relations (33.1) and (33.2) are obvious; to prove (33.3) one remarks that if $p = (\mathfrak{P}_1\mathfrak{P}_2)^2, \mathfrak{P}_1 \ne \mathfrak{P}_2$, $\mathfrak{P}_i \in S_o(K)$, then k_3 is the decomposition field for p and therefore $\kappa(p) = 1$. If $p = \mathfrak{P}^2$, $\mathfrak{P} \in S_o(K)$, then k_3 is the inertia subfield of K and therefore $\kappa(p) = -1$. Thus $\kappa(p) = 0$ if and only if $p = \mathfrak{P}^4$. Making use of (8) we get from (28):

$$\ell_{1p}(t)*\ell_{2p}(t) = (1-(\psi_1\psi_2)(p)t^2) \prod_{i,j=1}^{2} (1-\psi_1(\mathfrak{p}_i)\psi_2(\mathfrak{q}_j)t)^{-1} \tag{34}$$

when $p = \mathfrak{p}_1\mathfrak{p}_2 = \mathfrak{q}_1\mathfrak{q}_2$, $\mathfrak{p}_i \in S_o(k_1)$, $\mathfrak{q}_j \in S_o(k_2)$;

$$\ell_{1p}(t)*\ell_{2p}(t) = (1+(\psi_1\psi_2)(p)t^2)\prod_{i=1,2}(1-\psi_1(\mathfrak{p}_i^2)\psi_2(p)t^2)^{-1} \tag{35}$$

when $p = \mathfrak{p}_1\mathfrak{p}_2$, $\mathfrak{p}_i \in S_o(k_1)$, $p \in S_o(k_2)$. Finally if p remains prime in both k_1 and k_2, then it follows from (28) and (8) that

$$\ell_{1p}(t)*\ell_{2p}(t) = (1-(\psi_1\psi_2)(p)t^2)^{-1}. \tag{36}$$

On the other hand, since $G(K|k) \cong \mathbb{Z}/2\mathbb{Z} \times \mathbb{Z}/2\mathbb{Z}$, we have

$$p = \prod_{j=1}^{4}\mathfrak{P}_j,\ \mathfrak{P}_j \in S_o(K) \text{ when } p \text{ splits in both } k_1 \text{ and } k_2, \tag{37}$$

and

$$p = \mathfrak{P}_1\mathfrak{P}_2,\ \mathfrak{P}_i \in S_o(K) \text{ when } p \nmid D(k_i|k),\ i = 1,2, \text{ but } p \text{ remains prime in one of the fields } k_1,k_2 \text{ (or in both of them).} \tag{38}$$

Moreover,

$$\ell_{1p}(t)*\ell_{2p}(t) = \prod_{i=1}^{2}(1-\psi_1(\mathfrak{p}_i)\psi_2(\mathfrak{q})t)^{-1},\ p = (\mathfrak{P}_1\mathfrak{P}_2)^2,\ \mathfrak{P}_i \in S_o(K) \tag{39}$$

when $p = \mathfrak{p}_1\mathfrak{p}_2 = \mathfrak{q}^2$, $\mathfrak{p}_i \in S_o(k_1)$, $\mathfrak{q} \in S_o(k_2)$,

and

$$\ell_{1p}(t)*\ell_{2p}(t) = (1-\psi_1(p)\psi_2(\mathfrak{q}^2)t^2)^{-1},\ p = \mathfrak{P}^2,\ \mathfrak{P} \in S_o(K) \tag{40}$$

when $p \in S_o(k_1)$, $p = \mathfrak{q}^2$ with $\mathfrak{q} \in S_o(k_2)$. Here we have tacitly assumed that $\mathfrak{p}_1 \neq \mathfrak{p}_2$, $\mathfrak{q}_1 \neq \mathfrak{q}_2$, $\mathfrak{P}_i \neq \mathfrak{P}_j$ when $i \neq j$. Collecting the information about local factors from (32) - (40) one obtains (26). This completes the proof of proposition 2.

We evaluate now $L(s,\vec{\psi})$ expressing it in terms of the Euler product $L(s,\Phi)$, defined in §1, and L-functions Hecke.

<u>Lemma 4</u>. Let, in notations of (3) and (4),

$$\rho = \rho_1 \otimes \ldots \otimes \rho_r,\ \chi = \operatorname{tr}\rho. \tag{41}$$

Then

$$L(s,\vec{\psi}) = \prod_{p \in S_o} \ell_p(\vec{\chi}, |p|^{-s})^{-1} , \tag{42}$$

and

$$\ell_p(\vec{\chi},t) = \Phi_p(t)^{-1} \ell_p(\rho,t) \ell''_p(\vec{\chi},t)^{-1}, \tag{43}$$

where $\Phi_p(t)$ is given by (1.41) and (1.42); moreover

$$\ell''_p(\vec{\chi},t) = 1 \quad \text{for} \quad p \notin S_o(\vec{\chi}). \tag{44}$$

<u>Proof</u>. It follows from Proposition 1 and relations (1.31), (1.29), (1.36) and (1.40).

<u>Notations 1</u>. Let $K = k_1 \cdot \ldots \cdot k_r$ be the composite field of $k_1,\ldots,k_r$ and let E be the smallest Galois extension of k containing K.

<u>Proposition 3</u>. There are number fields E_j and grossencharacters λ_j such that

$$k \subseteq E_j \subseteq E , \quad \lambda_j \in gr(E_j), \quad 1 \leq j \leq a, \tag{45}$$

and that

$$\rho = \sum_{j=1}^{a} \oplus \operatorname{Ind}_{W_1(E|E_j)}^{W_1(E|k)} \lambda_j . \tag{46}$$

<u>Proof</u>. The assertion is obvious if $r = 1$. Suppose it holds for r-1 characters $\psi_1,\ldots,\psi_{r-1}$ and let

$$\rho' = \rho_1 \otimes \ldots \otimes \rho_{r-1} .$$

Then there are E'_j and λ'_j such that

$$k \leq E'_j \leq E , \quad \lambda'_j \in gr(E'_j), \quad 1 \leq j \leq a', \tag{47}$$

and

$$\rho' = \sum_{j=1}^{a'} \oplus \operatorname{Ind}_{W_1(E|E'_j)}^{W_1(E|k)} \lambda'_j . \tag{48}$$

We decompose $G(E|k)$ into double cosets:

$$G(E|k) = \bigcup_{g\in\Gamma_j} G(E|E'_j)g\, G(E|k_r), \quad 1 \leq j \leq a'$$

and choose for each g in Γ_j an element w_g in $W_1(E|k)$ whose image under the natural homomorphism of $W_1(E|k)$ on $G(E|k)$ coincides with g. This leads to a decomposition

$$W_1(E|k) = \bigcup_{g\in\Gamma_j} W_1(E|E'_j)w_g\, W_1(E|k_r).$$

Let $F_{jg} = E'^{g}_j\cdot k_r$ be the composite field of E'^{g}_j and k_r. We have

$$w_g^{-1}W_1(E|E'_j)w_g \cap W_1(E|F_{jg}) = W_1(E|F_{jg})$$

and

$$[W_1(E|k):W_1(E|F_{jg})] = [F_{jg}:k] < \infty \quad .$$

Therefore it follows from proposition I.2.2 that

$$(\mathrm{Ind}_{W_1(E|E'_j)}^{W_1(E|k)}\lambda'_j) \otimes \rho_r = \sum_{g\in\Gamma_j} \oplus\, \mathrm{Ind}_{W_1(E|F_{jg})}^{W_1(E|k)}\lambda'_{jg} \tag{49}$$

with

$$\lambda'_{jg}: \alpha \mapsto \psi_r(\alpha)\lambda'_j(w_g\alpha w_g^{-1}) \text{ for } \alpha \in w_g^{-1}W_1(E|E'_j)w_g \cap W_1(E|k_r).$$

By basic properties of Weil groups, λ'_{jg} may be regarded as a grossen-character in F_{jg}:

$$\lambda'_{jg}: \alpha \to \lambda'_j((N_{F_{jg}/E_{jg}}\alpha)^{g^{-1}})\psi_r(N_{F_{jg}/k_r}\alpha) \quad \text{for } \alpha \in C_{F_{jr}} \quad ,$$

or

$$\lambda'_{jg} = (\lambda^g_j\cdot N_{F_{jg}/E_{jg}})(\psi_r\cdot N_{F_{jg}/k_r}), \quad E_{jg} := E'^{g}_j \;. \tag{50}$$

One obtains (46) from (49) and (50):

$$\rho = \sum_{j=1}^{a'} \sum_{g\in\Gamma_j} \oplus\, \mathrm{Ind}_{W_1(E|F_{jg})}^{W_1(E|k)}\lambda'_{jg} \;, \quad \lambda'_{jg} \in \mathrm{gr}(F_{jg}). \tag{51}$$

This proves proposition 3 and provides an explicit construction for E_j and λ_j in (45) (cf. (50) and (51)).

Corollary 1. In notations (41), (45), we have

$$\ell_p(\rho,t) = \prod_{j=1}^{a} \prod_{\substack{\mathfrak{p}|p \\ \mathfrak{p}\in S_o(E_j)}} (1-\lambda_j(\mathfrak{p})t^{f(\mathfrak{p})}), \tag{52}$$

where $N_{E_j/k}\mathfrak{p} = p^{f(\mathfrak{p})}$.

Proof. Identity (52) follows from (46) and proposition I.3.4.

Corollary 2. Suppose $k_1,\dots,k_r$ are linearly disjoint over k, i.e. $[K:k] = d$, then $a = 1$ and $\lambda_1 = \psi$, where

$$\psi := \prod_{j=1}^{r} \psi_j \circ N_{K/k_j}, \quad \psi \in gr(K), \quad E_1 = K. \tag{53}$$

Proof. Let $H_i = W_1(E|k_i)$, $1 \le i \le r$. Then conditions of the Corollary I.2.1 are satisfied; moreover, $H^{(r)} = W_1(E|K)$ and $(\psi_i|H^{(r)}) = \psi_i \circ N_{K/k_i}$. Therefore the assertion follows from (I.2.13) and (I.3.12).

Lemma 5. The following relation holds:

$$S_o(\vec{\chi}) \subseteq \{p | p \in S_o(k),\ p|D(K|k) \vee \exists j(\psi_j(p) = 0)\}. \tag{54}$$

Proof. Let $p \in S_o(k)$ and suppose that $p \nmid D(k_j|k)$ and $\psi_j(p) \neq 0$. Then it follows from (I.3.11) and (I.3.12) that $\ell_p(\rho_j,t)$ is a polynomial of degree $d_j = \sum_{\mathfrak{p}|p} f(\mathfrak{p})$. Therefore, in view of (I.3.8), we conclude that $p \notin S_o(\rho_j)$. Thus

$$S_o(\rho_j) \subseteq \{p | p \in S_o(k),\ p|D(k_j|k) \vee \psi_j(p) = 0\}. \tag{55}$$

Relation (54) is a consequence of (55) and (1.33).

Notations 2. Let $\mathfrak{p} \in S_o(k_j)$ and $N_{k_j/k}\mathfrak{p} = p^{f(\mathfrak{p})}$, $p = \mathfrak{p}^{e(\mathfrak{p})}\mathfrak{q}$ with $\mathfrak{p} \nmid \mathfrak{q}$, so that $e(\mathfrak{p})$ and $f(\mathfrak{p})$ denote the ramification index and the inertia degree of $\mathfrak{p}$ in $k_j|k$ respectively; let $\vec{\mathfrak{p}} = (\mathfrak{p}_1,\dots,\mathfrak{p}_r)$, $\mathfrak{p}_j \in S_o(k_j)$, $\mathfrak{p}_j|p$, $1 \le j \le r$, for some p in $S_o(k)$. We write then $\vec{\mathfrak{p}}|p$ and denote by $F(\vec{\mathfrak{p}})$ the least common multiple of the relative degrees $f(\mathfrak{p}_j)$, $1 \le j \le r$; let

$$H(\vec{\mathfrak{p}}) = F(\vec{\mathfrak{p}})^{-1} \prod_{j=1}^{r} f(\mathfrak{p}_j), \quad g_j(\vec{\mathfrak{p}}) = f(\mathfrak{p}_j)^{-1} F(\vec{\mathfrak{p}}). \tag{56}$$

Finally, let

$$S_o(\vec{k}) = \{p \mid p \in S_o(k),\ \exists \vec{\mathfrak{p}}(\vec{\mathfrak{p}} \mid p \wedge \exists i,j((e(\mathfrak{p}_i), e(\mathfrak{p}_j)) > 1))\}. \tag{57}$$

Corollary 3. The set $S_o(\vec{k})$ is finite.

Proof. Let $p \in S_o(k)$. Then

$$\exists i(\mathfrak{p} \in S_o(k_i) \wedge \mathfrak{p} \mid p \wedge e(\mathfrak{p}) > 1).$$

Thus

$$S_o(\vec{k}) \subseteq \{p \mid p \in S_o(k),\ p \mid D(K|k)\}. \tag{58}$$

The following statement is elementary.

Lemma 6. Let F be the least common multiple of r natural numbers m_j, $1 \le j \le r$, and let $H = F^{-1} \prod_{j=1}^{r} m_j$. Let, in notations used in §1, $|(\alpha_1,\ldots,\alpha_r)| := \prod_{j=1}^{r} \alpha_j$ and let

$$A = \{(\varepsilon_1,\ldots,\varepsilon_r) \mid \varepsilon_j^{m_j} = 1,\ 1 \le j \le r\}.$$

Then

$$\prod_{\varepsilon \in A} (1 - |\varepsilon| u) = (1 - u^F)^H \quad \text{in } \mathbb{C}[u]. \tag{59}$$

Proof. Assuming that the lemma is true for $r-1$ integers $m_1,\ldots,m_{r-1}$ and denoting by F' the least common multiple of these integers, we get

$$\prod_{\varepsilon \in A} (1 - |\varepsilon| u) = \prod_{\varepsilon \in A_r} (1 - (\varepsilon u)^{F'})^{H'}, \tag{60}$$

where $H' = (\prod_{j=1}^{r-1} m_j) F'^{-1}$, $A_r = \{\varepsilon \mid \varepsilon^{m_r} = 1\}$. Let

$$F_r' = (F', m_r), \quad m_r' = m_r F_r'^{-1}.$$

It follows from (60) that

$$\prod_{\varepsilon \in A} (1-|\varepsilon|u) = (1-u^{F'm_r'})^{H'F_r'} .$$

On the other hand,

$$F'm_r' = F , \quad F_r'H' = H ,$$

and lemma follows by induction on r.

Lemma 7. Let, as in (1.30),

$$\ell_p'(\vec{\chi},t) = \det(1-t\rho_1(\sigma_p) \otimes \ldots \otimes \rho_r(\sigma_p)), \quad p \in S_o(k),$$

with ρ_j defined by

$$\rho_j = \mathrm{Ind}_{W(k_j)}^{W(k)} \psi_j , \quad 1 \leq j \leq r. \tag{61}$$

Then

$$\ell_p'(\vec{\chi},t) = \prod_{\vec{\mathfrak{p}}|p} (1-t^{F(\vec{\mathfrak{p}})} \prod_{j=1}^{r} \psi_j(\mathfrak{p}_j^{g_j(\vec{\mathfrak{p}})}))^{H(\vec{\mathfrak{p}})} , \tag{62}$$

where $\vec{\mathfrak{p}}$ ranges over r-tuples of prime divisors $(\mathfrak{p}_1,\ldots,\mathfrak{p}_r)$ such that $\mathfrak{p}_j|p$ in $S_o(k_j)$, $1 \leq j \leq r$.

Proof. It follows from (61) and (I.3.12) that

$$\ell_p(\rho_j,t) = \prod_{\substack{\mathfrak{p}|p \\ \mathfrak{p}\in S_o(k_j)}} (1-\psi_j(\mathfrak{p})t^{f(\mathfrak{p})}), \quad 1 \leq j \leq r. \tag{63}$$

In particular, $\ell_p(\rho_j,t)$ is of degree $\delta_i = \sum'_{\mathfrak{p}|p} f(\mathfrak{p})$, where $\sum'_{\mathfrak{p}|p}$ is extended over the finite set of primes

$$T_j(p) = \{\mathfrak{p} | \mathfrak{p} \in S_o(k_j), \ \mathfrak{p}|p, \ \psi_j(\mathfrak{p}) \neq 0\}.$$

For each $\mathfrak{p}$ in $T_j(p)$ let $\varphi_j(\mathfrak{p})$ be one of the roots of the equation $t^{f(\mathfrak{p})} = \psi_j(\mathfrak{p})$; let $\vec{\mathfrak{p}} = (\mathfrak{p}_1,\ldots,\mathfrak{p}_r)$, $\mathfrak{p}_j \in S_o(k_j)$ for $1 \leq j \leq r$ and let

$$A(\vec{\mathfrak{p}}) = \{(\varepsilon_1,\ldots,\varepsilon_r) \mid \varepsilon_j^{f(\mathfrak{p}_j)} = 1, \quad 1 \le j \le r\}.$$

It follows from (63) and Lemma 1.3 that, in notations of (1.27),

$$\ell_p(\vec{\chi},t) = \prod_{\vec{\mathfrak{p}}|p} \prod_{\alpha\in A(\vec{\mathfrak{p}})} (1-|\alpha|t \prod_{j=1}^{r} \varphi_j(\mathfrak{p}_j))\tilde{\Phi}_p(t)^{-1}, \tag{64}$$

where we set $\varphi_j(\mathfrak{p}) = 0$ when $\mathfrak{p}|p$, $\mathfrak{p} \in S_o(k_j)$, $\psi_j(\mathfrak{p}) = 0$ and where $\tilde{\Phi}_p(t)$ is a polynomial in $\mathbb{C}[t]$ of degree not higher than $\prod_{j=1}^{2} \delta_j - 1$.
By (60), (64) and (56), we get

$$\ell_p(\vec{\chi},t) = \tilde{\Phi}_p(t)^{-1} \prod_{\vec{\mathfrak{p}}|p} (1-t^{F(\vec{\mathfrak{p}})} \prod_{j=1}^{r} \psi_j(\mathfrak{p}_j^{g_j(\vec{\mathfrak{p}})}))^{H(\vec{\mathfrak{p}})} . \tag{65}$$

Identity (62) follows from (65) and (1.29); moreover, we see that $\tilde{\Phi}_p(t) = \Phi_p(t)$.

<u>Corollary 4</u>. The following identity holds:

$$\ell''_p(\vec{\chi},t) = \ell'_p(\vec{\chi},t)^{-1}\ell_p(\rho,t) \quad \text{for} \quad p \in S_o(\vec{\chi}), \tag{66}$$

where $\ell'_p(\vec{\chi},t)$ and $\ell_p(\rho,t)$ are given by (62) and (52), respectively.

<u>Proof</u>. It follows from (1.36), (52) and (62).

<u>Proposition 4</u>. The following identity holds:

$$L(s,\vec{\psi}) = L(s,\Phi)^{-1} \prod_{j=1}^{a} L(s,\lambda_j) \prod_{p\in S_o(\chi)} \ell''_p(\vec{\chi},|p|^{-s}), \quad s \in \mathbb{C}_+, \tag{67}$$

where $\ell''_p(\vec{\chi},t)$ is given by (66) and where λ_j, $1 \le j \le a$, is the grossencharacter constructed in the proposition 3. Moreover, the set $S_o(\vec{\chi})$ satisfies (55) and thereby it is determined in terms of $\vec{\psi}$ alone.

<u>Proof</u>. It follows from theorem 1 and relations (1.21), (42), (43), (55), (66).

<u>Notations 3</u>. Let $p \in S_o(k)$, $\vec{\mathfrak{p}} = (\mathfrak{p}_1,\ldots,\mathfrak{p}_r)$, $\mathfrak{p}_j \in S_o(k_j)$ for $1\le j\le r$ and suppose that $\vec{\mathfrak{p}}|p$. We define a finite subset of $S_o(K)$ as follows:

$$\mathcal{H}(\vec{\mathfrak{p}}) = \{\mathfrak{P} \mid \mathfrak{P} \in S_o(K),\ N_{K/k_j}\mathfrak{P} = \mathfrak{p}_j^{\,g_j(\vec{\mathfrak{p}})}\}. \tag{68}$$

Lemma 8. Suppose that the fields $k_1,\dots,k_r$ are linearly disjoint over k, that is $[K:k] = d$, and let $p \in S_o(k)\setminus S_o(\vec{k})$. Then

$$\bigcup_{\vec{\mathfrak{p}}\mid p} \mathcal{H}(\vec{\mathfrak{p}}) = \{\mathfrak{P} \mid \mathfrak{P} \in S_o(K),\ \mathfrak{P} \mid p\}, \tag{69}$$

$$\operatorname{card} \mathcal{H}(\vec{\mathfrak{p}}) = H(\vec{\mathfrak{p}}),\quad \vec{\mathfrak{p}} \mid p, \tag{70}$$

and

$$e(\mathfrak{P}) = \prod_{j=1}^{r} e(\mathfrak{p}_j) \quad \text{for} \quad \mathfrak{P} \in \mathcal{H}(\vec{\mathfrak{p}}),\quad \vec{\mathfrak{p}} \mid p, \tag{71}$$

where $\vec{\mathfrak{p}} = (\mathfrak{p}_1,\dots,\mathfrak{p}_r)$ and $e(\mathfrak{P})$ denotes the ramification index of $\mathfrak{P}$ in $K|k$.

Proof. If $r = 1$, then $K = k_1$ and the assertion is obvious. Let $r = 2$ and let $p \in S_o(k)\setminus S_o(\vec{k})$; then

$$(e(\mathfrak{p}_1), e(\mathfrak{p}_2)) = 1 \quad \text{whenever} \quad (\mathfrak{p}_1,\mathfrak{p}_2) \mid p. \tag{72}$$

Let

$$K(p) = K \otimes_k k_p,\quad k_j(p) = k_j \otimes_k k_p,\quad K_j(\mathfrak{p}) = K \otimes_{k_j} k_{j\mathfrak{p}}, \tag{73}$$

where $p \in S_o(k)$, $\mathfrak{p} \in S_o(k_j)$, $j = 1,2$. Since k_1 and k_2 are linearly disjoint over k, we have an identity

$$K = k_1 \otimes_k k_2 . \tag{74}$$

On the other hand, the decomposition law for prime ideals gives (cf., e.g., [17], p. 76):

$$K(p) = \sum_{\mathfrak{P}\mid p} \oplus K_{\mathfrak{P}},\quad k_j(p) = \sum_{\mathfrak{p}\mid p} k_{j\mathfrak{p}},\quad j = 1,2, \tag{75}$$

where $\mathfrak{P}$ and $\mathfrak{p}$ range over $S_o(K)$ and $S_o(k_j)$, respectively. Relations (73) - (75) show that

$$k_{1\mathfrak{P}_1} \underset{k_p}{\otimes} k_{2\mathfrak{P}_2} \subseteq k_1(p) \underset{k_p}{\otimes} k_{2\mathfrak{P}_2} = K_2(\mathfrak{P}_2),$$

and analogously

$$k_{1\mathfrak{P}_1} \underset{k_p}{\otimes} k_{2\mathfrak{P}_2} \subseteq K_1(\mathfrak{P}_1).$$

Thus

$$k_{1\mathfrak{P}_1} \underset{k_p}{\otimes} k_{2\mathfrak{P}_2} \leq K_1(\mathfrak{P}_1) \cap K_2(\mathfrak{P}_2) \tag{76}$$

whenever $(\mathfrak{P}_1,\mathfrak{P}_2)|p$. By (75),

$$\sum_{\vec{\mathfrak{P}}|p} k_{1\mathfrak{P}_1} \underset{k_p}{\otimes} k_{2\mathfrak{P}_2} = k_1(p) \underset{k_p}{\otimes} k_2(p), \quad \vec{\mathfrak{P}} := (\mathfrak{P}_1,\mathfrak{P}_2), \tag{77}$$

and since

$$K_j(\mathfrak{P}) = \sum_{\mathfrak{p}|\mathfrak{P}} \oplus K_{\mathfrak{p}}, \quad j = 1,2, \quad \mathfrak{p} \in S_o(K),$$

it follows also that

$$\sum_{\vec{\mathfrak{P}}|p} \oplus (K_1(\mathfrak{P}_1) \cap K_2(\mathfrak{P}_2)) = K(p). \tag{78}$$

By (74),

$$K(p) = k_1(p) \underset{k_p}{\otimes} k_2(p). \tag{79}$$

Since $k_{1\mathfrak{P}_1} \underset{k_p}{\otimes} k_{2\mathfrak{P}_2}$ and $K_1(\mathfrak{P}_1) \cap K_2(\mathfrak{P}_2)$ are finite dimensional k_p-algebras, one concludes from (76) - (79) that

$$k_{1\mathfrak{P}_1} \underset{k_p}{\otimes} k_{2\mathfrak{P}_2} = K_1(\mathfrak{P}_1) \cap K_2(\mathfrak{P}_2). \tag{80}$$

Let $K_{\mathfrak{P}_1,\mathfrak{P}_2} = k_{1\mathfrak{P}_1} \cdot k_{2\mathfrak{P}_2}$ be the composite field of $k_{1\mathfrak{P}_1}$ and $k_{2\mathfrak{P}_2}$ (in a fixed algebraic closure of k_p). It follows from (72) that

$$[K_{\mathfrak{P}_1,\mathfrak{P}_2}:k_p] = F(\vec{\mathfrak{P}})e(\mathfrak{P}_1)e(\mathfrak{P}_2), \quad \vec{\mathfrak{P}} := (\mathfrak{P}_1,\mathfrak{P}_2), \tag{81}$$

and therefore

$$k_1\mathfrak{p}_1 \underset{k_p}{\otimes} k_2\mathfrak{p}_2 = \sum_{\mathfrak{P}\in\mathcal{H}(\vec{\mathfrak{p}})} \oplus K_{\mathfrak{P}}, \quad K_{\mathfrak{P}} \cong K_{\mathfrak{p}_1,\mathfrak{p}_2}. \tag{82}$$

The statement of lemma 8 for $r = 2$ follows from (81) and (82). Assuming that $r \geq 2$ we proceed now by induction on r. Suppose the assertion is valid for $k_1,\ldots,k_{r-1}$ and let K_{r-1} denote the composite of these fields. Choose $\vec{\mathfrak{p}} = (\mathfrak{p}_1,\ldots,\mathfrak{p}_r)$ with $\vec{\mathfrak{p}}\,|\,p$ and let

$$\mathcal{H}_o(\vec{\mathfrak{p}}) = \{\mathfrak{P}\mid \mathfrak{P}\in S_o(K_{r-1}),\ N_{K_{r-1}/k_i}\mathfrak{P}=\mathfrak{p}_i^{\,g_o(\mathfrak{p}_i)},$$

$$1 \leq i \leq r-1\},$$

where

$$g_o(\mathfrak{p}_i) = f(\mathfrak{p}_i)^{-1}F_o(\vec{\mathfrak{p}}), \quad 1 \leq i \leq r-1,$$

and $F_o(\vec{\mathfrak{p}})$ denotes the least common multiple of $f(\mathfrak{p}_1),\ldots,f(\mathfrak{p}_{r-1})$. It follows from the inductive assumption that

$$\operatorname{card}\mathcal{H}_o(\vec{\mathfrak{p}}) = F_o(\vec{\mathfrak{p}})^{-1}\prod_{i=1}^{r-1} f(\mathfrak{p}_i). \tag{83}$$

Applying the assertion of the lemma to two fields K_{r-1} and k_r one obtains an equation:

$$\operatorname{card}\{\mathfrak{P}\mid\mathfrak{P}\in S_o(K), N_{K/k_r}\mathfrak{P}=\mathfrak{p}_r^{\,g(\mathfrak{p}_r)},\ N_{K/K_{r-1}}\mathfrak{P}=\mathfrak{P}_o^{\,g_o(\vec{\mathfrak{p}})}\} = F_1(\vec{\mathfrak{p}}), \tag{84}$$

where $\mathfrak{P}_o \in \mathcal{H}_o(\vec{\mathfrak{p}})$, $\mathfrak{p}_r \in S_o(k_r)$, $g_o(\vec{\mathfrak{p}}) = F(\vec{\mathfrak{p}})F_o(\vec{\mathfrak{p}})^{-1}$, and $F_1(\vec{\mathfrak{p}}) = F_o(\vec{\mathfrak{p}})f(\mathfrak{p}_r)F(\vec{\mathfrak{p}})^{-1}$. Equation (70) follows from (83) and (84). Relations (69) and (71) are simple consequences of the inductive assumption and the assertion of the lemma applied to K_{r-1} and k_r.

<u>Corollary 5</u>. Let $[K:k] = d$ and let $p \in S_o(k)\setminus S_o(\vec{k})$. Then

$$\ell_p'(\vec{X},t) = \prod_{\mathfrak{P}|p}(1-\psi(\mathfrak{P})t^{f(\mathfrak{P})}), \tag{85}$$

where $\mathfrak{P}$ ranges over $S_o(K)$, $N_{K/k}\mathfrak{P}= p^{f(\mathfrak{P})}$ for $\mathfrak{P}|p$, and ψ is defined by (53).

Proof. It follows from (62) and (68) - (70).

Proposition 5. Suppose that $[K:k] = d$, then

$$L(s,\vec{\psi}) = L(s,\Phi)^{-1}L(s,\psi) \prod_{p\in S_o(\vec{k})} \ell''_p(\vec{\chi},|p|^{-s}) \quad \text{for} \quad s \in \mathbb{C}_+ , \tag{86}$$

where

$$\ell''_p(\vec{\chi},t) = \prod_{\mathfrak{P}|p} (1-\psi(\mathfrak{P})t^{f(\mathfrak{P})})\ell'_p(\vec{\chi},t)^{-1} \tag{87}$$

with $\ell'_p(\vec{\chi},t)$ given by (62).

Proof. Identity (86) follows from (67), (85) and Corollary 2. Relation (87) is a consequence of (66) and Corollary 2.

Corollary 6. If $[K:k] = d$ and $S_o(\vec{k}) = \emptyset$, then

$$L(s,\vec{\psi}) = L(s,\Phi)^{-1}L(s,\psi) \quad \text{for} \quad s \in \mathbb{C}_+. \tag{88}$$

Proof. Immediate from (86).

§6. Proof of the theorem 4.2.

Let $E|k$ be a finite Galois extension with r_1 real and r_2 complex places and let $m+1$ denote its degree:

$$m+1 = r_2+2r_2 = [E:\mathbb{Q}]. \tag{1}$$

Let $\rho \in R(E|k)$ and let

$$\rho|_{C_E} = \psi_1 \oplus \dots \oplus \psi_\ell, \ \psi_j \in gr(E), \ 1 \leq j \leq \ell, \ \ell = \dim \rho. \tag{2}$$

In notations of (I.5.29), let

$$v(\rho) = \max_{1\leq j\leq \ell} b(\psi_j) \tag{3}$$

and let

$$\mathfrak{f}(\rho) = \ell.c.m. \ (\mathfrak{f}(\psi_1,\dots,\mathfrak{f}(\psi_\ell)), \tag{4}$$

where $\mathfrak{f}(\psi_j)$ denotes the conductor of ψ_j. By (2) and (3),

$$v(\rho_1 \oplus \rho_2) = \max\{v(\rho_1), v(\rho_2)\}, \ \rho_i \in R(E|k), \ i = 1,2. \tag{5}$$

Let, for a subextension $E'|k$ of $E|k$,

$$\rho = \mathrm{Ind}_{W(E|E')}^{W(E|k)} \rho', \quad \chi = \mathrm{tr}\, \rho, \quad \chi' = \mathrm{tr}\, \rho',$$

and let

$$G(E|k) = \bigcup_{\gamma\in\Gamma} G(E|E')\gamma,$$

where Γ is a set of representatives of $G(E|k)$ modulo $G(E|E')$. Then

$$\chi(\alpha) = \sum_{\gamma\in\Gamma} \chi'(\gamma\alpha), \quad \alpha \in C_E. \tag{6}$$

Since

$$b(\psi^\gamma) = b(\psi), \quad \psi \in gr(E), \ \gamma \in G(E\ k),$$

where ψ^γ is defined by the equation

$$\psi^{\gamma}(\alpha) = \psi(\gamma\alpha), \qquad \alpha \in C_E ,$$

it follows from (2), (3) and (6) that

$$v(\rho) = v(\rho'). \tag{7}$$

Proposition 1. Let $\mathfrak{a} \in I_o(E)$ and let

$$\mathfrak{R} = \{\rho \mid \rho \in R(E|k), \ \mathfrak{f}(\rho) \mid \mathfrak{a}\} .$$

There is a real number $\alpha(\mathfrak{a})$ such that for each ρ in $\mathfrak{R}$ we have

$$\sum_{|p|<x} \chi(p) = g(\chi) \int_2^x \frac{du}{\log u} + O(x \exp(-\alpha(\mathfrak{a}) \frac{\log x}{\sqrt{\log x} + \log v(\rho)})),$$

$$\alpha(\mathfrak{a}) > 0, \tag{8}$$

where p ranges over the primes in $S_o(k)$,

$$\chi(p) := \operatorname{tr} \rho(\sigma_p)|V_p ,$$

and $g(\chi)$ denotes the multiplicity of the identical representation in ρ. Here the constant implied by O-symbol doesn't depend on x and ρ (but may depend on $\mathfrak{a}$).

Proof. We proceed by induction on $\ell := \dim \rho$. If $\ell = 1$, then $\chi \in gr(k)$ and therefore (8) follows from the estimate obtained in the theorem I.5.1. Suppose that ρ is reducible, say $\rho = \rho_1 \oplus \rho_2$, $\dim \rho_i < \ell$, $i = 1,2$. Then we deduce (8) from the corresponding estimates for ρ_1 and ρ_2 taking into account relation (5). If ρ is induced by ρ', say

$$\rho = \operatorname{Ind}_{W(E|E')}^{W(E|k)} \rho', \quad \rho' \in R(E'|k), \quad \chi' = \operatorname{tr} \rho' ,$$

then, by (I.3.25), we have

$$L(s,\chi) = L(s,\chi'). \tag{9}$$

Taking the logarithmic derivatives in (9) one obtains an identity:

$$\sum_{\substack{|\mathfrak{p}|=p \\ \mathfrak{p}\in S_o(k)}} \chi(\mathfrak{p}) = \sum_{\substack{|\mathfrak{p}'|=p \\ \mathfrak{p}'\in S_o(E')}} \chi'(\mathfrak{p}'), \qquad p \in S_o(\mathbb{Q}). \tag{10}$$

Relation (10) gives:

$$\sum_{|\mathfrak{p}|<x} \chi(\mathfrak{p}) = \sum_{|\mathfrak{p}'|<x} \chi'(\mathfrak{p}') + O(x^{1/2}), \tag{11}$$

where $\mathfrak{p}$ and $\mathfrak{p}'$ range over the primes in $S_o(k)$ and $S_o(E')$, respectively. Assuming that $\dim \rho' < \ell$ we deduce (8) from (11), (7) and the inductive assumption. In view of Corollary I.3.1, it remains to consider the primitive representations. Suppose that ρ is primitive and let $\rho \in \mathfrak{N}$. By the proposition I.3.5,

$$\rho = \rho_1 \otimes \rho_2, \ \rho_1 \in gr(k), \quad \rho_2 \in R_o(k). \tag{12}$$

Since $\rho \in \mathfrak{N}$, it follows from (12) that $\mathfrak{f}(\rho_2) | \mathfrak{a}$. Therefore

$$\rho_2 \in R_o(E_{\mathfrak{a}}|k), \ \rho \in R(E_{\mathfrak{a}}|k), \tag{13}$$

where $E_{\mathfrak{a}}$ is the ray class field modulo $\mathfrak{a}$ (say, ramified at all the infinite places of E). By the theorem I.2.1, there are φ_j, $1 \leq j \leq \nu$, such that

$$\rho_2 = \sum_{j=1}^{\nu} \oplus\, e_j \operatorname{Ind}_{W(E_{\mathfrak{a}}|k_j)}^{W(E_{\mathfrak{a}}|k)} \varphi_j \ , \ \varphi_j \in gr(k_j), \ e_j \in \{-1,1\}, \tag{14}$$

where k_j, $1 \leq j \leq \nu$, is an intermediate field, so that

$$E_{\mathfrak{a}} \supseteq k_j \supseteq k,$$

and φ_j may be regarded, via the class field theory, as one-dimensional representation of $G(E_{\mathfrak{a}}|k)$. In particular,

$$V(\varphi_j) = O(1), \tag{15}$$

where the O-constant depends on E and $\mathfrak{a}$ only. Relations (12), (14) and (I.3.12) give:

$$\chi_2(p) = \sum_{j=1}^{\nu} e_j \sum_{\mathfrak{p}|p}{}' \varphi_j(\mathfrak{p}), \quad p \in S_o(k), \quad \chi_2 := \mathrm{tr}\ \rho_2,$$

where the sum $\sum'_{\mathfrak{p}|p}$ is extended over primes $\mathfrak{p}$ in $S_o(k_j)$ satisfying the condition $N_{k_j/k}\mathfrak{p} = p$. Therefore

$$\chi(p) = \sum_{j=1}^{\nu} e_j \sum_{\mathfrak{p}|p}{}' \psi_j(\mathfrak{p})\varphi_j(\mathfrak{p}), \quad p \in S_o(k), \quad \psi_j = \rho_1 \cdot N_{k_j/k}. \tag{16}$$

In view of the estimate (I.5.45), it follows from (16) and (13) that

$$\sum_{|p|<x} \chi(p) = \sum_{j=1}^{\nu} e_j \sum_{|\mathfrak{p}|<x} \psi_j(\mathfrak{p})\varphi_j(\mathfrak{p}) + O(x^{1/2}) \tag{17}$$

with an O-constant depending on E only. We get from (17) and theorem I.5.1:

$$\sum_{|p|<x} \chi(p) = \sum_{j=1}^{\nu} e_j g(\psi_j)g(\varphi_j) \int_2^x \frac{du}{\log u} + O(\sum_{j=1}^{\nu} x^{\beta_j}) +$$

$$+ O(x \sum_{j=1}^{\nu} \exp(-c_8 \frac{\log x}{\sqrt{n_j \log x} + \log(a(\psi_j\varphi_j)b(\psi_j\varphi_j))})), \tag{18}$$

where $n_j = [k_j:\mathbb{Q}]$ and β_j denotes the possible exceptional zero of $L(s,\psi_j\varphi_j)$. It follows from (18) and (I.5.41) that

$$g(\chi) = \sum_{j=1}^{\nu} e_j g(\psi_j)g(\varphi_j). \tag{19}$$

Moreover,

$$b(\psi_j) \le v(\rho_j)^{c_1(\mathfrak{a})}, \quad 1 \le j \le \nu, \tag{20}$$

with $c_1(\mathfrak{a})$ depending on $E_{\mathfrak{a}}$ only. Finally $L(s,\psi_j\varphi_j)$ may have an exceptional zero only if $(\psi_j\varphi_j)^2 = 1$ and there are only O(1) (with an $\mathfrak{a}$-dependent O-constant) such possibilities. Therefore (8) follows from (18) - (20) and (15).

We shall make use of the following simple observation.

Lemma 1. Let χ be a simple character of a compact group G and let μ denote the Haar measure on G normalised by the condition $\mu(G) = 1$.

Then

$$\int_G \chi(t^{-1}h_1th_2^{-1})d\mu(t) = \chi(h_1)\overline{\chi(h_2)}\chi(1)^{-1} \text{ for } h_1\in G,\ h_2\in G. \tag{21}$$

<u>Proof</u>. Write $\chi = \operatorname{tr}\rho$, $\rho(h) = (a_{ij}(h))$, $h \in G$, $1 \le i, j \le \chi(1)$. Since G is compact, we may assume that ρ is an unitary representation. Then

$$\chi(t^{-1}h_1th_2^{-1}) = \operatorname{tr}(\rho(t)^{-1}\rho(h_1)\rho(t)\rho(h_2)^{-1}) =$$

$$= \sum_{i,j,\ell,m} \overline{a_{ji}(t)}a_{j\ell}(h_1)a_{\ell m}(t)\overline{a_{im}(h_2)}. \tag{22}$$

By the orthogonality relations (ρ is irreducible!),

$$\int_G d\mu(t)\overline{a_{ji}(t)}a_{\ell m}(t) = \delta_{j\ell}\delta_{im}\chi(1)^{-1}. \tag{23}$$

Equation (21) follows from (22) and (23).

Let $\mathfrak{a} \in I_o(k)$ and let

$$\mathfrak{g}(\mathfrak{a}) = \{\psi \mid \psi \in \operatorname{gr}(E),\ \mathfrak{f}(\psi) \mid \mathfrak{a}\}$$

be the group of all the grossencharacters in E whose conductor divides $\mathfrak{a}$ (we regard $\mathfrak{a}$ as a $G(E|k)$-invariant ideal in $I_o(E)$). By the proposition I.1.1., one can write

$$\mathfrak{g}(\mathfrak{a}) = \mathfrak{g}_1(\mathfrak{a}) \times \mathfrak{g}_2(\mathfrak{a}),\quad \mathfrak{g}_1(\mathfrak{a}) \cong \mathbb{Z}^m, \tag{24}$$

and $\mathfrak{g}_2(\mathfrak{a})$ is a finite abelian group. We fix a system of generators $\{\lambda_j \mid 1 \le j \le m\}$ of $\mathfrak{g}_1(\mathfrak{a})$ and define m functions

$$\varphi_j\colon C_E \to [-\tfrac{1}{2},\tfrac{1}{2}],\quad 1 \le j \le m,$$

subject to the conditions:

$$\lambda_j(\alpha) = \exp(2\pi i\varphi_j(\alpha)),\ -\tfrac{1}{2} \le \varphi_j(\alpha) < \tfrac{1}{2},\ \alpha \in C_E.$$

For each ε in $\mathbb{R}_+$ let

$$V(\varepsilon,\mathfrak{A}) = \{\alpha \mid \alpha \in C_E, \lambda(\gamma\alpha) = 1 \text{ for } \lambda \in \mathcal{G}_2(\mathfrak{A}),\ \gamma \in G(E|k),$$

$$|\varphi_j(\gamma\alpha)| < \tfrac{\varepsilon}{2} \text{ for } 1 \le j \le m,\ \gamma \in G(E|k)\}.$$

Further, for each p in $S_o(k)$ we choose an element τ_p in σ_p and consider a set

$$\mathcal{A}(g,\varepsilon,x;t) = \{p \mid p \in S_o(k), |p| < x, t^{-1}\tau_p t \in V(\varepsilon,\mathfrak{A})g\},$$

where t and g range over $W(E|k)$. Let

$$A(g,\varepsilon,x;t) = \operatorname{card} \mathcal{A}(g,\varepsilon,x;t)$$

and let

$$A_o(g,\varepsilon,x) = \int_{W_1(E|k)} A(g,\varepsilon,x;t)\,d\mu(t), \tag{25}$$

where μ denotes the Haar measure on $W_1(E|k)$ normalised by the condition

$$\mu(W_1(E|k)) = 1.$$

<u>Proposition 2</u>. The function

$$t \mapsto A(g,\varepsilon,x;t)$$

is measurable, so that $A_o(g,\varepsilon,x)$ is well defined. Moreover, there are two positive constants c_2 and c_3 such that

$$A_o(g,\varepsilon,x) = c_2\varepsilon^m \int_2^x \frac{du}{\log u} + O(x\exp(-c_3\sqrt{\log x})) \tag{26}$$

for any ε in the interval $0 < \varepsilon < 1$ and any x in $\mathbb{R}_+$. Here the O-constant, c_2 and c_3 may depend on $\mathfrak{A}$ but do not depend on g,ε,x.

<u>Proof</u>. Let

$$0 < \Delta < \frac{\varepsilon}{2} < \frac{\varepsilon}{2} + \Delta < \frac{1}{2}.$$

By lemma I.7.1, one can construct two functions $f_{\pm}: \mathbb{R} \to \mathbb{R}$ with the following properties:

$$f_+(x) = \begin{cases} 1 & \text{when } |x| < \frac{\varepsilon}{2} \\ 0 & \text{when } \Delta+\frac{\varepsilon}{2} \leq |x| \leq \frac{1}{2} \end{cases}, \qquad (27.1)$$

$$f_-(x) = \begin{cases} 1 & \text{when } |x| < \frac{\varepsilon}{2} - \Delta \\ 0 & \text{when } \frac{\varepsilon}{2} \leq |x| \leq \frac{1}{2} \end{cases}, \qquad (27.2)$$

and

$$f_{\pm} \in C^{\infty}(\mathbb{R}); \; f_{\pm}(x+1) = f_{\pm}(x), \; 0 \leq f_{\pm}(x) \leq 1 \quad \text{for} \quad x \in \mathbb{R}. \qquad (28)$$

Moreover,

$$f_{\pm}(x) = \sum_{\ell=-\infty}^{\infty} a_{\pm}(\ell)\exp(2\pi i\ell x), \quad x \in \mathbb{R}, \qquad (29)$$

and

$$a_{\pm}(0) = \varepsilon+O(\Delta), \quad |a_{\pm}(\ell)| \leq C_{\nu}(\Delta|\ell|)^{-\nu}, \; \nu \in \mathbb{N}, \qquad (30)$$

with C_{ν} depending on ν only (but not on the other parameters of the problem: ε,Δ,ℓ). Let

$$\tilde{h}_{\pm}(\alpha) = \frac{1}{|\mathfrak{g}_2(\mathfrak{a})|} \sum_{\lambda\in\mathfrak{g}_2(\mathfrak{a})} \lambda(\alpha) \prod_{j=1}^{m} \prod_{\gamma\in G(E|k)} f_{\pm}(\varphi_j(\gamma\alpha)) \qquad (31)$$

for $\alpha \in C_E$. Since $\mathfrak{a}$ is $G(E|k)$-invariant, it follows from (27), (31) and the definition of $V(\varepsilon,\mathfrak{a})$ that

$$\tilde{h}_+(\alpha) = 1 \quad \text{when} \quad \alpha \in V(\varepsilon,\mathfrak{a}), \qquad (32.1)$$

and

$$\tilde{h}_-(\alpha) = 0 \quad \text{when} \quad \alpha \notin V(\varepsilon,\mathfrak{a}). \qquad (32.2)$$

Moreover,

$$\tilde{h}_{\pm}(\alpha) \in \mathbb{R} \quad \text{for each} \quad \alpha \quad \text{in} \quad C_E. \tag{33}$$

Relations (29) and (31) give:

$$\tilde{h}_{+}(\alpha) = \sum_{\ell,\lambda} \tilde{a}_{\pm}(\ell)\psi_{\lambda}^{\ell}(\alpha), \tag{34}$$

where ℓ ranges over all the functions of the shape:

$$\ell : \{j \mid 1 \le j \le m\} \times G(E|k) \to \mathbb{Z},$$

and

$$\tilde{a}_{\pm}(\ell) := \frac{1}{|\mathfrak{g}_2(\mathfrak{a})|} \prod_{j=1}^{m} \prod_{\gamma \in G(E|k)} a_{\pm}(\ell(j,\gamma)), \tag{35}$$

while λ ranges over $\mathfrak{g}_2(\mathfrak{a})$; the character ψ_{λ}^{ℓ} is defined by the equation:

$$\psi_{\lambda}^{\ell}(\alpha) = \lambda(\alpha) \prod_{j=1}^{m} \prod_{\gamma \in G(E|k)} \lambda_j(\gamma\alpha)^{\ell(j,\gamma)}, \quad \alpha \in C_E, \tag{36}$$

so that

$$\psi_{\lambda}^{\ell} \in gr(E).$$

Let

$$\rho_{\lambda}^{\ell} = \mathrm{Ind}_{C_E}^{W(E|k)} \psi_{\lambda}^{\ell}, \quad \chi_{\lambda}^{\ell} = \mathrm{tr}\, \rho_{\lambda}^{\ell}, \tag{37}$$

and let

$$h_{\pm} = \frac{1}{[E:k]} \sum_{\ell,\lambda} \tilde{a}_{\pm}(\ell)\chi_{\lambda}^{\ell}, \quad h_{\pm} : W(E|k) \to \mathbb{C}. \tag{38}$$

It follows from (37) and (I.2.2) that

$$\chi_{\lambda}^{\ell}(\alpha) = \begin{cases} 0, & \alpha \notin C_E \\ \sum_{\gamma \in G(E|k)} \psi_{\lambda}^{\ell}(\gamma\alpha) & \text{for } \alpha \in C_E \end{cases}. \tag{39}$$

Relations (38), (39) and (34) show that

$$h_{\pm}(\alpha) = 0 \quad \text{for} \quad \alpha \notin C_E \tag{40.1}$$

and

$$h_{\pm}(\alpha) = \sum_{\gamma}^{*} \tilde{h}_{\pm}(\gamma\alpha) \quad \text{for} \quad \alpha \in C_E, \tag{40.2}$$

where $\sum\limits_{\gamma}^{*}$ stands for $\frac{1}{[E:k]} \sum\limits_{\gamma\in G(E|k)}$. In particular, it follows from (40) that actually

$$h_{\pm}(\alpha) \in \mathbb{R} \quad \text{for} \quad \alpha \in W(E|k). \tag{41}$$

In view of (31), we have

$$\tilde{h}_{\pm}(\gamma\alpha) = \tilde{h}_{\pm}(\alpha) \quad \text{for} \quad \alpha \in W(E|k),\ \gamma \in G(E|k).$$

Therefore (40) may be rewritten as follows:

$$h_{\pm}(\alpha) = \begin{cases} 0 & , \quad \alpha \notin C_E \\ \tilde{h}_{\pm}(\alpha), & \alpha \in C_E \end{cases} . \tag{42}$$

Let

$$A_{\pm}(t) = \sum_{\substack{|p|<x \\ p\in S_o(k)}} h_{\pm}(t^{-1}\tau_p t g^{-1}), \qquad t \in W(E|k), \tag{43}$$

and let

$$A_{\pm}^{o} = \int_{W_1(E|k)} d\mu(t) A_{\pm}(t). \tag{44}$$

Relations (32) and (41) - (44) give:

$$A_{-}^{o} \leq A_o(g,\varepsilon,x) \leq A_{+}^{o}. \tag{45}$$

On the other hand, let

$$\chi_{\lambda}^{\ell} = \sum_{i} \varphi_{i;\lambda,\ell} \quad , \tag{46}$$

where $\varphi_{i;\lambda,\ell}$ is a simple character of $W_1(E|k)$. By lemma 1, we get from (38), (43), (44) and (46):

$$A_{\pm} = \sum_{\ell,\lambda,i} c_{\pm}(i;\lambda,\ell) \sum_{|p|<x} \varphi_{i;\lambda,\ell}(\tau_p), \tag{47}$$

where

$$c_{\pm}(i;\lambda,\ell) = \frac{1}{[E:k]} \tilde{a}_{\pm}(\ell) \overline{\varphi_{i;\lambda,\ell}(g)} \varphi_{i;\lambda,\ell}(1)^{-1}. \tag{48}$$

It follows from (37) that

$$\rho_{\lambda}^{\ell}|C_E = \sum_{\gamma\in G(E|k)} \oplus (\psi_{\lambda}^{\ell})^{\gamma}, \tag{49}$$

where we write, for brevity,

$$\varphi^{\gamma}(\alpha) = \varphi(\gamma\alpha) \quad \text{for } \gamma \in G(E|k),\ \alpha \in C_E,\ \varphi \in gr(E).$$

Thus $\varphi^{\gamma} \in gr(E)$ when $\varphi \in gr(E)$ and $\gamma \in G(E|k)$. Relation (49) shows at once that

$$v(\rho_{\lambda}^{\ell}) = b(\psi_{\lambda}^{\ell}) \tag{50}$$

in view of (3), and that

$$S_o(\rho_{\lambda}^{\ell}) \subseteq \{p | p \in S_o(k),\ p \nmid \mathfrak{a}D(E|k)\}, \tag{51}$$

cf. the proof of proposition I.3.3. It follows from (51) and (49) that

$$\varphi_{i;\lambda,\ell}(\tau_p) = \varphi_{i;\lambda,\ell}(\sigma_p) \quad \text{for } p \nmid \mathfrak{n}D(E|k). \tag{52}$$

Relations (47), (50), (51) and (8) give:

$$A_{\pm}^{o} = \sum_{\ell,\lambda,i} c_{\pm}(i;\lambda,\ell) g(\varphi_{i;\lambda,\ell}) \int_2^x \frac{du}{\log u} + R(x) \tag{53}$$

with

$$R(x) = O(x \sum_{\ell,\lambda,i} |c_{\pm}(i;\lambda,\ell)| \exp(-\alpha(\mathfrak{a}) \frac{\log x}{\sqrt{\log x} + \log b(\psi_{\lambda}^{\ell})})). \tag{54}$$

It follows from (49) that if ρ_{λ}^{ℓ} contains the identical representation, then $\psi_\lambda^\ell = 1$. Moreover, if $\psi_\lambda^\ell = 1$ then it follows from the definition (37) that ρ contains the identical representation with multiplicity one. Therefore relations (53) and (48) give:

$$A_\pm^o = \frac{1}{[E:k]} \sum_{\ell,\lambda}^{*} \tilde{a}_\pm(\ell) \int_2^x \frac{du}{\log u} + R(x), \tag{55}$$

where $\sum_{\ell,\lambda}$ denotes the sum extended over those ℓ and λ for which $\psi_\lambda^\ell = 1$. Let now

$$\lambda_j^\gamma = \prod_{i=1}^{m} \lambda_i^{e(j,\gamma;i)}, \quad e(j,\gamma;i) \in \mathbb{Z}. \tag{56}$$

It follows from (36) that $\psi_\lambda^\ell = 1$ if and only if

$$\lambda = 1 \text{ and } \sum_{j=1}^{m} \sum_{\gamma\in G(E|k)} \ell(j,\gamma)e(j,\gamma;i) = 0, \quad 1 \leq i \leq m. \tag{57}$$

We get from (57):

$$\sum_{\ell,\lambda}^{*} \tilde{a}_\pm(\ell) = \sum_{\ell} \tilde{a}_\pm(\ell) \int_{\mathfrak{b}} du \exp(2\pi i \sum_{j,j'=1}^{m} \sum_{\gamma\in G(E|k)} e(j,\gamma;j')\ell(j,\gamma)u_{j'}) \tag{58}$$

where the integral is extended over the cube

$$\mathfrak{b} = \{u | u \in \mathbb{R}^m, |u_j| \leq \tfrac{1}{2} \text{ for } 1 \leq j \leq m\}.$$

Since the series obtained after substitution of (35) in (58) converges absolutely (and uniformly in the cube $\mathfrak{b}$), we can sum up this series before integration over $\mathfrak{b}$. In view of (29), this procedure leads to the equation

$$\sum_{\ell,\lambda}^{*} \tilde{a}_\pm(\ell) = \frac{1}{|\mathfrak{g}_2(\mathfrak{a})|} \int_{\mathfrak{b}} du \prod_{j=1}^{m} \prod_{\gamma\in G(E|k)} f_\pm(\sum_{i=1}^{m} u_i e(j,\gamma;i))du.$$

It follows from (27) and (28) that this equation may be rewritten as follows:

$$\sum_{\ell,\lambda}^{*} \tilde{a}_\pm(\lambda) = \frac{1}{|\mathfrak{g}_2(\mathfrak{a})|} \int_{\mathfrak{b}(\varepsilon)} du + O(\Delta), \tag{59}$$

where

$$\mathscr{b}(\varepsilon) := \{u \mid u \in b, \left|\sum_{i=1}^{m} u_i e(j,\gamma;i)\right| < \frac{\varepsilon}{2} \quad \text{for} \quad \gamma \in G(E|k),\ 1 \le j \le m\},$$

and where the O-constant depends on $\mathfrak{a}$ (but not on ε and Δ). In view of (56), we have

$$e(j,1;i) = \delta_{ji}. \tag{60}$$

Since $0 < \varepsilon < 1$, relation (60) shows that

$$\mathscr{b}(\varepsilon) = \{u \mid u \in \mathbb{R}^m, \left|\sum_{i=1}^{m} u_i e(j,\gamma;i)\right| < \frac{\varepsilon}{2} \quad \text{for} \quad \gamma \in G(E|k),\ 1 \le j \le m\}.$$

Thus (59) gives:

$$\sum_{\ell,\lambda}^{*} \tilde{a}_{\pm}(\lambda) = c_4\varepsilon^m + O(\Delta), \tag{61}$$

where

$$c_4 = \frac{1}{|\mathscr{g}_2(\mathfrak{a})|} \int_{\mathscr{b}(1)} du. \tag{62}$$

Since $\mathscr{b}(1)$ is an open subset of $\mathbb{R}^m$ and $\mathscr{b}(1) \neq \emptyset$ (because it contains the origin), we get from (62) an inequality:

$$c_4 > 0. \tag{63}$$

Relations (55), (61) and (63) give:

$$A_{\pm}^{\bullet} = c_2\varepsilon^m \int_2^x \frac{du}{\log u} + R(x) + O(x\Delta);$$

therefore it follows from (45) that

$$A_o(g,\varepsilon,x) = c_2\varepsilon^m \int_2^x \frac{du}{\log u} + R(x) + O(x\Delta). \tag{64}$$

To estimate $R(x)$ let us remark that, by (36),

$$\log b(\psi_\lambda^\ell) = O(\log \|\ell\|), \tag{65}$$

where

$$\|\ell\| := \sum_{j=1}^{m} \prod_{\gamma \in G(E|k)} (1+|\ell(j,\gamma)|). \tag{66}$$

Since, by (30), (35), and (48),

$$c_{\pm}(i,\lambda;\ell) = O(\Delta^{-\nu\mu}\|\ell\|^{-\nu} C_\nu), \ \nu \in \mathbb{N}, \tag{67}$$

with $\mu := (m+1)[E:k]$, it follows from (54), (65), and (66) that

$$R(x) = O(x \sum_{\ell} \|\ell\|^{-\nu} \exp(-\alpha(\mathfrak{a}) \frac{\log x}{\sqrt{\log x} + \log\|\ell\|}) C_\nu \Delta^{-\nu\mu}). \tag{68}$$

We choose now $\nu = 3$ in the relation (67) and remark that

$$\text{card}\{\ell \mid \|\ell\| = L\} = O_\varepsilon(L^\varepsilon) \quad \text{for} \quad \varepsilon > 0.$$

Then (68) gives

$$R(x) = O(x \sum_{L=1}^{\infty} L^{-2} \exp(-\alpha(\mathfrak{a}) \frac{\log x}{\sqrt{\log x} + \log L}) \Delta^{-3\mu}). \tag{69}$$

Relation (26) follows from (64) and (69) when one adjusts Δ properly. This completes the proof of proposition 2.

We return now to notations 4.1 and suppose that $\mathfrak{M}$ is a <u>finite</u> subset of $R(k)$. Let $E|k$ be a finite Galois extension for which

$$\mathfrak{M} \subseteq R(E|k). \tag{70}$$

Since $\mathfrak{M}$ is finite, one can choose an ideal $\mathfrak{a}$ in $I_o(k)$ which satisfies the following condition:

$$\mathfrak{f}(\rho) \mid \mathfrak{a} \text{ for each } \rho \text{ in } \mathfrak{M}. \tag{71}$$

<u>Lemma 2</u>. Let $\mathfrak{M}$ be a finite set satisfying relations (70) and (71). Then there is a <u>positive</u> real number L such that

$$P_{\mathfrak{M}}(g,\varepsilon;x_1,x_2) \geq A(g,\varepsilon L^{-1},x_2;t) - A(g,\varepsilon L^{-1},x_1;t) - |S_o(\mathfrak{M})|, \tag{72}$$

as soon as

$$0 < \varepsilon < 1, \quad x_1 > x_2 \geq 2, \quad g \in W(k), \quad \text{and} \quad t \in W(E|k). \tag{73}$$

Thus L depends on $\mathfrak{M}$ but not on $\varepsilon, x_1, x_2, g, t$.

Proof. Let $\rho \in \mathfrak{M}$. By (70), we can write

$$\rho|_{C_E} = \psi_1^\rho \oplus \ldots \oplus \psi_{\nu(\rho)}^\rho, \quad \psi_j^\rho \in gr(E), \quad 1 \leq j \leq \nu(\rho), \tag{74}$$

where $\nu(\rho) = \dim \rho$. Since ρ is equivalent to an unitary representation, one may set

$$\rho(g) = (a_{ij}^\rho(g)), \quad 1 \leq i, j \leq \nu(\rho), \quad g \in W(k),$$

where $(a_{ij}^\rho(g))$ is an unitary matrix and, in particular,

$$|a_{ij}^\rho(g)| \leq 1 \quad \text{for} \quad \rho \in \mathfrak{M}, \; g \in W(k), \quad 1 \leq i,j \leq \nu(\rho). \tag{75}$$

Let

$$p \in \mathfrak{A}(g,\varepsilon,x;t) \setminus S_o(\rho), \quad \tau_p \in \sigma_p, \quad t \in W(E|k). \tag{76}$$

Since $p \notin S_o(\rho)$, we have

$$\chi(\sigma_p) = \chi(t^{-1}\tau_p t) \text{ for } t \in W(E|k), \quad \chi := \operatorname{tr} \rho. \tag{77}$$

Thus, by (77),

$$|\chi(\sigma_p) - \chi(g)| = |\operatorname{tr}((\rho(t^{-1}\tau_p t g^{-1}) - 1)\rho(g))|. \tag{78}$$

By definition of $\mathfrak{A}(g,\varepsilon,x;t)$, we have

$$t^{-1}\tau_p t g^{-1} \in C_E. \tag{79}$$

Relations (74), (75), (78), and (79) give:

$$|\chi(\sigma_p) - \chi(g)| \leq \sum_{j=1}^{\nu(\rho)} |\psi_j^\rho(t^{-1}\tau_p t g^{-1}) - 1|. \tag{80}$$

Moreover, it follows from (4) and (71) that

$$\psi_j^\rho \in \mathfrak{G}(\mathfrak{A}), \quad 1 \leq j \leq \nu(\rho),$$

and therefore

$$\psi_j^\rho = \mu_{j\rho} \sum_{i=1}^{m} \lambda_i^{\ell(i;j,\rho)} \quad \text{with } \mu_{j\rho} \in \mathcal{G}_2(\mathfrak{a}),\ \ell(i;j,\rho) \in \mathbb{Z}. \tag{81}$$

For $p \in \mathcal{A}(g,\varepsilon,x;t)$ we have $\mu_{j\rho}(t^{-1}\tau_p tg^{-1}) = 1$; thus, by (80) and (81),

$$|\chi(\sigma_p)-\chi(g)| \leq \sum_{j=1}^{\nu(\rho)} |1- \prod_{i=1}^{m} \lambda_i^{\ell(i;j,\rho)}(t^{-1}\tau_p tg^{-1})|,$$

or

$$|\chi(\sigma_p)-\chi(g)| \leq \sum_{j=1}^{\nu(\rho)} |1-\exp(2\pi i \sum_{a=1}^{m} \varphi_a(t^{-1}\tau_p tg^{-1})\ell(a;j,\rho))|. \tag{82}$$

Let

$$L = \max_{\rho\in\mathfrak{M}} \{(2\pi) \sum_{j=1}^{\nu(\rho)} \sum_{a=1}^{m} |\ell(a;j,\rho)|\}$$

and let

$$0 < \varepsilon_1 L < 1.$$

Then it follows from (82) that

$$|\chi(\sigma_p)-\chi(g)| < \varepsilon_1 L \quad \text{for } p \in \mathcal{A}(g,\varepsilon_1,x;t)\setminus S_0(\rho). \tag{83}$$

Relations (83) and (4.2) give:

$$\mathcal{A}(g,\varepsilon L^{-1},x;t)\setminus S_0(\mathfrak{M}) \subseteq \mathcal{P}_{\mathfrak{M}}(g,\varepsilon) \quad \text{for } t \in W(E|k). \tag{84}$$

Estimate (72) is an immediate consequence of (84) and the definitions (4.3), (4.4). This proves the lemma.

<u>Proof of theorem 4.2</u>. Relations (72) and (25) give:

$$P_{\mathfrak{M}}(g,\varepsilon;x_1,x_2) \geq A_0(g,\varepsilon L^{-1},x_2) - A_0(g,\varepsilon L^{-1},x_1) - |S_0(\mathfrak{M})|, \tag{85}$$

where L doesn't depend on ε,g,x_1,x_2 (but may depend on $\mathfrak{M}$). Estimate (4.5) follows from (85) and (26). This proves theorem 4.2.

<u>Conjecture 1</u>. Let, in notations 4.1, $0 < \varepsilon < 1$ and let

$$Q_{\mathfrak{M}}(g,\varepsilon) = \{h | h \in W_1(E|k),\ |\chi(h)-\chi(g)| < \varepsilon \quad \text{for} \quad \chi \in \check{\mathfrak{M}}\},$$

where $\mathfrak{M}$ is a <u>finite</u> subset of $R(E|k)$. Then

$$P_{\mathfrak{M}}(g,\varepsilon;x_1,x_2) = \mu(Q_{\mathfrak{M}}(g,\varepsilon)) \int_{x_1}^{x_2} \frac{du}{\log u} + O(x_2 \exp(-c_3 \sqrt{\log x_2})), \tag{C 86}$$

where c_3 and the implied by the O-symbol constant do not depend on ε, g, x_1, x_2 (but may depend on $\mathfrak{M}$) and where $c_3 > 0$.

<u>Corollary 1</u>. There are c_1 and c_2 such that

$$\mu(Q_{\mathfrak{M}}(g,\varepsilon)) \geq c_1 \varepsilon^{c_2}, \quad c_1 > 0, \quad c_2 > 0. \tag{C 87}$$

<u>Proof</u>. Relation (87) is an immediate consequence of (86) and (4.5).

Chapter III. Ideals with equal norms and integral points on non-form varieties.

§1. On character sums extended over ideals having equal norms.

Let, in notations of (II.1.26),

$$a(\mathfrak{m},\vec{\chi}) = \prod_{j=1}^{r} a(\mathfrak{m},\chi_j), \quad \mathfrak{m}\in I_o(k), \quad \chi_j := \operatorname{tr}\rho_j , \tag{1}$$

so that

$$L(s,\vec{\chi}) = \sum_{\mathfrak{m}\in I_o(k)} a(\mathfrak{m},\vec{\chi})\,|\mathfrak{m}|^{-s}. \tag{2}$$

Let

$$A(x,\vec{\chi}) = \sum_{|\mathfrak{m}|<x} a(\mathfrak{m},\vec{\chi}) \tag{3}$$

and let

$$\Pi(x,\vec{\chi}) = \sum_{|p|<x} a(p,\vec{\chi}), \tag{4}$$

where $\mathfrak{m}$ and p range over $I_o(k)$ and $S_o(k)$, respectively. Further let, as in §II.1,

$$\rho = \rho_1 \otimes \ldots \otimes \rho_r , \quad \chi := \operatorname{tr}\rho \;;\; d = \prod_{j=1}^{r} d_j,\; d_j := \dim\rho_j. \tag{5}$$

We are concerned with asymptotic, as $x\to\infty$, evaluation of $A(x,\vec{\chi})$ and $\Pi(x,\vec{\chi})$.

__Lemma 1.__ In notations of (4), (5) and (II.1.33), we have:

$$\Pi(x,\vec{\chi}) = \sum_{|p|<x} \chi(p) + O(d\,|S_o(\vec{\chi})|). \tag{6}$$

__Proof.__ It follows from (2) and (II.1.28) - (II.1.31) that

$$a(p,\vec{\chi}) = \operatorname{tr}(\rho_1(\sigma_p) \otimes \ldots \otimes \rho_r(\sigma_p)) \text{ for } p\in S_o(k). \tag{7}$$

By (7) and (II.1.33),

$$a(p,\vec{\chi}) = \chi(p) \quad \text{for} \quad p \in S_o(k) \setminus S_o(\vec{\chi}). \tag{8}$$

Since

$$\dim \rho_j(\sigma_p) \leq d_j \quad , \quad 1 \leq j \leq r,$$

estimate (6) follows from (7) and (8).

Remark 1. Lemma 1 and theorem I.5.2 settle the question of estimating $\Pi(x,\vec{\chi})$.

Notations 1. By theorem I.3.1, the function $L(s,\chi)$ can be decomposed into product (I.3.26) of abelian L-functions. We fix one of such representations and let, in notations of (I.3.26), $N = \sum_{j=1}^{n} n_j$, $n_j := [k_j:\mathbb{Q}]$. Suppose, furthermore, that

$$e_j = \begin{cases} 1 & \text{when} \quad 1 \leq j \leq m' \\ -1 & \text{when} \quad j \geq m'+1 \end{cases} .$$

and let

$$N_1 = \sum_{j=1}^{m'} n_j, \quad N_2 = \sum_{j=m'+1}^{m} n_j \ , \ N = N_1+N_2, \ N_3 = \sum_{j=1}^{m} n_j^2. \tag{9}$$

Theorem 1. The following estimates hold:

$$A(x,\vec{\chi}) = xP(\vec{\chi},\log x) + R(\vec{\chi},x), \tag{10}$$

where

$$R(\vec{\chi},x) = O(C(\vec{\chi})x \exp(- \frac{c_1 \sqrt{\log x}}{N})), \quad c_1 > 0, \tag{11}$$

and where

$$P(\vec{\chi},t) = \sum_{j=0}^{g(\chi)-1} t^j a_j(\vec{\chi}) \quad \text{when} \quad g(\chi) \geq 1,$$

$$P(\vec{\chi},t) = 0 \quad \text{when} \quad g(\chi) = 0.$$

The constant $C(\vec{\chi})$ can be explicitly evaluated in terms of $\vec{\chi}$; the coefficients $a_j(\vec{\chi})$ are exactly expressible in terms of $L(s,\vec{\chi})$ and its derivatives, in particular, $P(\vec{\chi},t)$ is equal to the residue of $L(s,\vec{\chi})$ at $s = 1$ when $g(\chi) = 1$. Moreover,

$$R(\vec{\chi},x) = O(C_1(\varepsilon;n\delta_1)B(\chi)x^{1-\frac{2}{N_1+2}+\varepsilon}(\log x)^{nd}), \quad \varepsilon > 0, \tag{AW 12}$$

if ρ is of AW type. Here $B(\chi)$ is defined as in (I.4.34), while $C_1(\varepsilon;n\delta_1)$ is an exactly computable constant; $\delta_1 := d|S_o(\vec{\chi})|$. If ρ is of Lindelöf type, then

$$R(\chi,x) = O(C_2(\varepsilon;N_3,\delta_1)(a(\chi)b(\chi))^{\varepsilon}x^{1/2+\varepsilon}), \quad \varepsilon > 0, \tag{L 13}$$

with an exactly computable $C_2(\varepsilon;N_3,\delta_1)$ (cf. (I.6.13)).

<u>Lemma 2</u>. If $\operatorname{Re} s > \frac{1}{2}$, then

$$|L(s,\vec{\chi})| \leq c_2 e^{\delta_1}(2\sigma-1)^{-\delta}|L(s,\chi)|, \quad \operatorname{Re} s =: \sigma , \tag{14}$$

where $\delta := nd^{d+1}$.

<u>Proof</u>. By theorem II.1.1,

$$|L(s,\vec{\chi})| = |L(s,\chi)|\cdot|L_o(s,\vec{\chi})^{-1}| \prod_p |\Phi_p(|p|^{-s})|. \tag{15}$$

It follows from (II.1.37) and (II.1.38) that

$$|L_o(s,\vec{\chi})| \leq e^{\delta_1} \quad \text{for } \operatorname{Re} s \geq \frac{1}{2} . \tag{16}$$

Relations (II.1.41) and (II.1.42) give:

$$|\Phi_p(|p|^{-s})| \leq 1 + \sum_{m=2}^{d-1} p^{-\sigma}|b_m(p)|,$$

and

$$|b_m(p)| \leq (m+1)d^m, \qquad 2 \leq m \leq d-1,$$

therefore

$$|\Phi_p(|p|^{-s})| \le 1+|p|^{-2\sigma}d^{d+1}. \qquad (17)$$

By (17),

$$\prod_p |\Phi_p(|p|^{-s})| \le \zeta_k(2\sigma)^{d^{d+1}} \qquad (18)$$

Estimate (14) follows from (16), (18) and (I.4.5), (I.4.6).

Lemma 3. In notations I.4.1,

$$L(s,\vec{\chi}) \ll \zeta_k(s)^{d_1} * \ldots * \zeta_k(s)^{d_r}. \qquad (19)$$

Proof. It follows from the definition (II.1.26) and relation (I.4.4).

Proposition 1. If ρ is of AW type (respectively Lindelöf type), then relations (10) and (12) (respectively (13)) hold true.

Proof. In view of (19), relation (I.4.33) gives:

$$A(x,\vec{\chi}) = \frac{1}{2\pi i}\int_{1+\eta-iT}^{1+\eta+iT} \frac{x^w}{w} L(w,\vec{\chi})\,dw + O\Big(\frac{x^{1+\eta}}{T\eta^{nd}}\Big) + O\Big(C_3(\varepsilon)\frac{x^{1+\varepsilon}}{T}\log x\Big),$$

where $C_3(\varepsilon)$ is defined so that

$$\zeta_{k_1}(s) * \ldots * \zeta_{k_r}(s) << \sum_{m=1}^{\infty} C_3(\varepsilon) m^{\varepsilon} m^{-s}, \ \varepsilon > 0.$$

We take $\eta = (\log x)^{-1}$, move the contour of integration to the line $\mathrm{Re}\, s = \frac{1}{2}+\varepsilon$ and apply estimates (14) and (I.4.17) (respectively (I.6.12)). This leads to (10) and (12) (respectively (13)).

We need a few results concerning the coefficients of a Dirichlet series. Let $f(s)$ be a function meromorphic in the half-plane

$$\mathfrak{B} = \{u+it \mid u \ge 1 - \frac{c_3}{\log(b_1(2+|t|)^n)}, \ t \in \mathbb{R}\},$$

$$b_1 \ge 1 \ge c_3 > 0,$$

and suppose that

$$f(s) = \sum_{m=1}^{\infty} a_m m^{-s} \quad \text{for} \quad \text{Re } s > 1, \tag{20.1}$$

and

$$\sum_{m=1}^{\infty} |a_m| m^{-\sigma} < \infty \qquad \text{for} \quad \sigma > 1. \tag{20.2}$$

Let, moreover,

$$f(u+it) = O(b_2(2+|t|)^{\gamma}) \quad \text{for} \quad |t| \geq c_4, \quad u+it \in \mathcal{B} \tag{21}$$

with $0 < \gamma < 1$, $b_2 \geq 1$. Finally suppose that f is regular in

$\mathcal{B} \setminus \{\alpha_j | 1 \leq j \leq \nu\}$, $\alpha_j > 1-c_3 \log(b_1 2^n)$, $1 \leq j \leq \nu$,

and that

$$\lim_{s \to \rho_j} (s-\alpha_j)^{g_j} f(s) \neq 0, \infty \quad \text{for} \quad g_j \in \mathbb{Z}, \quad 1 \leq j \leq \nu.$$

Let $A(x) = \sum_{m<x} a_m$.

Lemma 4. There is a numerical constant c_4 such that

$$A(x) = \sum_{j=1}^{\nu} x^{\alpha_j} P_j(\log x) + O(b_1 b_2 x \exp(-c_5 \frac{\sqrt{\log x}}{n})) + O(\sum_{x \leq m < x\beta} |a_m|,) \tag{22}$$

where $\beta = 1 + \exp(-c_5 \frac{\sqrt{\log x}}{n})$, $c_5 > 0$, and $P_j = 0$ when $g_j \leq 0$; if $g_j \geq 1$, then $P_j(t)$ is a polynomial of degree $g_j - 1$ exactly computable in terms of f and its derivatives at $s = \alpha_j$, $1 \leq j \leq \nu$. The implied by the O-symbol constant can be evaluated in terms of $\sup|h(s)|$ for $s \in \mathcal{B}$, $|\text{Im } s| \leq c_4$; $h(s) := \prod_{j=1}^{\nu} (s-\alpha_j)^{g_j} f(s)$.

Proof. Let

$$A_1(x) = \sum_{m<x} a_m \log(xm^{-1}).$$

Since

$$\frac{1}{2\pi i}\int_{\alpha-i\infty}^{\alpha+i\infty}\frac{y^s}{s^2}\,ds = \begin{cases} 0 & \text{for } y<1 \\ \log y & \text{for } y>1 \end{cases} \qquad \text{when } \alpha>1,\ y>0,$$

we have an identity

$$A_1(x) = \frac{1}{2\pi i}\int_{\alpha-i\infty}^{\alpha+i\infty}\frac{x^s}{s^2}f(s)\,ds. \tag{23}$$

Moving contour of integration to the line $\mathrm{Re}\, s = 1-\dfrac{c_3}{\log(b_1(2+|t|)^n)}$,

where $t := \mathrm{Im}\, s$, one obtains from (20), (21) and (23):

$$A_1(x) = \sum_{j=1}^{\nu} x^{\alpha_j}\tilde{P}_j(\log x) + O(b_2 x I(x)), \tag{24}$$

where

$$I(x) = \int_{-\infty}^{\infty}(2+|t|)^{\gamma}(1+|t|^2)^{-1}\exp\Bigl(-\frac{c_3\log x}{\log(b_1(2+|t|)^n)}\Bigr)dt. \tag{25}$$

Choose t_o to satisfy the conditions

$$\log(b_1(1+t_o)^n) = \sqrt{\log x}, \qquad t_o\in\mathbb{R}_+.$$

Then

$$I_1 = \int_{-t_o}^{t_o}\frac{(2+|t|)^{\gamma}}{1+|t|^2}\exp\Bigl(-\frac{c_3\log x}{\log(b_1(2+|t|)^n)}\Bigr)dt = O(e^{-c_3\sqrt{\log x}}), \tag{26}$$

and

$$I_2 = \int_{|t|\geq t_o}\frac{(2+|t|)^{\gamma}}{1+|t|^2}\exp\Bigl(-\frac{c_3\log x}{\log(b_1(2+|t|)^n)}\Bigr)dt = O\Bigl(b_1\exp\Bigl(-(1-\gamma)\frac{\sqrt{\log x}}{n}\Bigr)\Bigr). \tag{27}$$

Since $I(x) = I_1+I_2$, we deduce from (24) - (27) an estimate:

$$A_1(x) = \sum_{j=1}^{\nu} x^{\alpha_j}\tilde{P}_j(\log x) + O\Bigl(b_1 b_2 x\exp\Bigl(-\frac{(1-\gamma)c_3\sqrt{\log x}}{n}\Bigr)\Bigr). \tag{28}$$

If $1\leq\beta<2$, then

$$A_1(\beta x) = \sum_{m<x\beta} a_m \log \frac{x\beta}{m} = A(x) \log\beta + A_1(x) + \sum_{x\leq m<\beta x} a_m \log \frac{x\beta}{m} . \qquad (29)$$

Therefore

$$A(x) = (A_1(\beta x)-A_1(x))(\log \beta)^{-1} + O(\sum_{x\leq m\leq \beta x} |a_m|). \qquad (30)$$

Taking $\beta = 1 + \exp(-\frac{c_5\sqrt{\log x}}{n})$, $c_5 = (1-\gamma)c_3$, we deduce (22) from (28) and (30).

Lemma 5. If, in conditions of lemma 4, we have $a_m \geq 0$ for each m, then

$$A(x) = \sum_{j=1}^{\nu} x^{\alpha_j} P_j(\log x) + O(b_1 b_2 x \exp(-c_5 \frac{\sqrt{\log x}}{n})), \quad c_5 > 0. \qquad (31)$$

Proof. Since $a_m \geq 0$ for each m, it follows from (29) that

$$A(x) \leq \frac{A_1(\beta x)-A_1(x)}{\log \beta} . \qquad (32)$$

On the other hand,

$$A_1(\frac{x}{\beta}) = \sum_{m<x\beta^{-1}} a_m \log \frac{x}{m\beta} = A_1(x) - A(x)\log \beta + \sum_{x\leq m\beta<\beta x} a_m \log \frac{\beta m}{x} ,$$

so that

$$A(x) \geq \frac{A_1(x)-A_1(x\beta^{-1})}{\log \beta} . \qquad (33)$$

Relation (31) follows from (28), (32) and (33).

Proposition 2. Relations (10) and (11) of theorem 1 are valid.

Proof. Let

$$\varphi(t,\chi) = \max_{1\leq j\leq m} \varphi(t,\chi_j)$$

with $\varphi(t,\chi_j)$ defined as in §I.5, and let

$$D = \prod_{j=1}^{m} D_j ,$$

where D_j denotes the discriminant of k_j. Relation (I.5.19) shows that

$$L(s,\vec{\chi}) = O(C_4(\vec{\chi})(2+|t|)^{1/2}) \text{ for } |t| \geq 1, \quad s \in \mathfrak{B}_1, \tag{34}$$

where

$$\mathfrak{B}_1 = \{u+it \mid u \geq 1-c_6(\log \varphi(t,\chi))^{-1}, \quad t \in \mathbb{R}\}.$$

Write

$$\zeta_{k_1}(s)^{d_1} * \ldots * \zeta_{k_r}(s)^{d_r} = \sum_{m=1}^{\infty} a_m m^{-s}. \tag{35}$$

The function (35) may be regarded as a scalar product of L-functions $L(s,\lambda_j)$, $1 \leq j \leq r$, where $\lambda_j = d_j$. Since

$$dI = d_1 I \otimes \ldots \otimes d_r I ,$$

where I denotes the identical representation of $W(k)$, is of AW type, it follows from Proposition 1 that

$$\sum_{m<x} a_m = xP_0(\log x) + O(C_1(\varepsilon;nd)B(d)x^{1-\frac{2}{N+2}+\varepsilon}(\log x)^{nd}), \tag{36}$$

where $P_0(t)$ is a polynomial of degree $d-1$. The assertion of proposition 2 follows from (34), (14), (36), (19) and Lemma 4.

<u>Proof of theorem 1</u>. It follows from Proposition 1 and Proposition 2.

<u>Remark 2.</u> We haven't used lemma 5; it is included for its own sake only

<u>Remark 3</u>. It is known classically, [40], [41], that

$$\sum_{\substack{|\mathfrak{a}|<x \\ \mathfrak{a}\in I_0(k)}} \chi(\mathfrak{a}) = g(\chi)x(k)x + O(x^{\gamma}C_5(\chi)) \quad \text{for } \chi \in gr(k) \tag{37}$$

with $\gamma \le \frac{n-1}{n+1}$. Moreover, (37) implies (cf. [3], p. 128) that

$$\gamma > \frac{1}{2} - \frac{1}{2n} \quad . \tag{38}$$

One should think that, in general,

$$A(x,\chi) = xP(\chi,\log x) + O(C_6(\chi)x^{1/2-\alpha}), \quad \alpha > 0, \qquad \text{(Co 39)}$$

with $P(\chi,t)$ defined as in (I.4.35), $\rho \in R(k)$, $\chi := \operatorname{tr} \rho$. However, it would be a rather bold conjecture because (39) is easily seen to imply the Artin-Weil conjecture but doesn't seem to be a direct consequence of it.

<u>Notations 2.</u> Let

$$I_o = \{\vec{\mathfrak{a}} \mid \vec{\mathfrak{a}} = (\mathfrak{a}_1,\ldots,\mathfrak{a}_r),\ \mathfrak{a}_j \in I_o(k_j),\ N_{k_1/k}\mathfrak{a}_1 = \ldots = N_{k_r/k}\mathfrak{a}_r\}$$

and let

$$S_o = \{\vec{\mathfrak{p}} \mid \vec{\mathfrak{p}} \in I_o,\ \vec{\mathfrak{p}} = (\mathfrak{p}_1,\ldots,\mathfrak{p}_r),\ \mathfrak{p}_j \in S_o(k_j) \text{ for } 1 \le j \le r\}.$$

Write

$$|\vec{\mathfrak{a}}| = N_{k_1/\mathbb{Q}}\mathfrak{a}_1 \quad \text{for} \quad \vec{\mathfrak{a}} \in I_o \quad ,$$

and let

$$\vec{\chi}(\vec{\mathfrak{a}}) = \prod_{j=1}^{r} \chi_j(\mathfrak{a}_j) \text{ for } \vec{\mathfrak{a}} \in I_o,\ \vec{\chi} = (\chi_1,\ldots,\chi_r),\ \chi_j \in gr(k_j).$$

Finally let

$$Gr = gr(k_1)\times\ldots\times gr(k_r),$$

and let

$$A(\vec{\psi},x) = \sum_{\substack{|\vec{\mathfrak{a}}|<x \\ \in I_o}} \vec{\psi}(\vec{\mathfrak{a}}), \quad \vec{\psi} \in Gr .$$

<u>Proposition 3.</u> Let $\vec{\psi} \in Gr$ and let

$$\rho_j = \mathrm{Ind}_{W(k_j)}^{W(k)} \psi_j, \quad \chi_j := \mathrm{tr}\, \rho_j, \quad 1 \leq j \leq r.$$

Then

$$A(\vec{\psi},x) = A(\vec{\chi},x), \tag{40}$$

and the representation

$$\rho = \rho_1 \otimes \ldots \otimes \rho_r$$

is of AW type.

Proof. Equation (40) follows from (II.5.4). Since ρ_j, $1 \leq j \leq r$, is monomial, ρ is of AW type by proposition I.2.2.

Remark 4. It follows from proposition 3 that estimate (12) holds for $R(\vec{\chi},x)$ when $\vec{\chi}$ is defined as in (40); moreover, explicit calculations of §II.5 allow for a more precise evaluation of $P(\vec{\chi},t)$ in (10) than it is possible in general.

Notation 3. Let

$$\Pi_o(x,\vec{\psi}) = \sum_{\substack{|\vec{\mathfrak{p}}|<x \\ \vec{\mathfrak{p}} \in S_o}} \vec{\psi}(\vec{\mathfrak{p}}) \qquad \text{for } \vec{\psi} \in Gr .$$

Lemma 6. Let $\vec{\psi} \in Gr$ and let $\vec{\chi}$ be defined as in (40). Then

$$\Pi_o(x,\vec{\psi}) = \Pi(x,\vec{\chi}) + O(d\sqrt{x}). \tag{41}$$

Proof. It follows from the definitions that

$$\Pi_o(x,\vec{\psi}) = \Pi(x,\vec{\chi}) + \sum_{|p|<x^{1/2}} \sum_{\ell=2}^{d} \sum_{\substack{\vec{\mathfrak{p}} \in S_o \\ N_{k_1/\mathbb{Q}}\mathfrak{p}_1 = p^\ell}} \vec{\psi}(\vec{\mathfrak{p}}) . \tag{42}$$

Relation (41) is an immediate consequence of (42).

§2. Equidistribution of ideals with equal norms.

Let K be the composite field of k_j, $1 \le j \le r$. We introduce the following assumption:

$$k = \mathbb{Q};\ k_j|k \ \text{ is a Galois extension for } 1 \le j \le r;\ [K:k] = d, \tag{1}$$

where, as always,

$$[k_j:k] =: d_j \quad \text{and} \quad d := \prod_{j=1}^{r} d_j . \tag{2}$$

Let $S_o(\vec{k})$ be defined as in (II.5.57).

<u>Definition 1.</u> We say that the fields $k_1,\ldots,k_r$ are <u>arithmetically independent</u> if (1) holds and $S_o(\vec{k}) = \emptyset$.

<u>Notations 1.</u> We keep notations 1.2 and let

$$\mathfrak{T} = \mathfrak{T}_1\times\ldots\times\mathfrak{T}_r, \quad H = H_1\times\ldots\times H_r, \quad I = I(k_1)\times\ldots\times I(k_r),$$

where $\mathfrak{T}_j$ and H_j, $1 \le j \le r$, denote the real (d_j-1)-dimensional torus assigned to k_j in §I.1 and the ideal class group of k_j, respectively. Let the homomorphism

$$f\colon I \to \mathfrak{T}$$

be defined by the equation

$$f(\vec{\mathfrak{a}}) = (f_1(\mathfrak{a}_1),\ldots,f_r(\mathfrak{a}_r)) \text{ for } \vec{\mathfrak{a}} \in I,\ \vec{\mathfrak{a}} = (\mathfrak{a}_1,\ldots,\mathfrak{a}_r),$$

where the homomorphism $f_j\colon I(k_j) \to \mathfrak{T}_j$, $1 \le j \le r$, is defined by (I.1.4). Let H_K denote the ideal class group of K and let

$$N\colon H_K \to H, \quad N\colon A \mapsto (N_{K/k_1}A,\ldots,N_{K/k_r}A), \tag{3}$$

be a homomorphism of H_K in H. Furthermore, for $\vec{A} \in H$ and $\vec{\chi} \in \hat{H}$ we write

$$\vec{\chi}(\vec{A}) = \prod_{j=1}^{r} \chi_j(A_j) \text{ when } \vec{A} = (A_1,\ldots,A_r),\ \vec{\chi} = (\chi_1,\ldots,\chi_r).$$

For $\tau \subseteq \mathcal{T}$, $\vec{A} \in H$, $x \in \mathbb{R}_+$ let

$$\pi(\tau,\vec{A};x) = \text{card}\,\{\vec{\mathfrak{p}} \mid \vec{\mathfrak{p}} \in S_o \cap \vec{A},\ f(\vec{\mathfrak{p}}) \in \tau,\ |\vec{\mathfrak{p}}| < x\}$$

and let

$$\iota(\tau,\vec{A};x) = \text{card}\,\{\vec{\mathfrak{a}} \mid \vec{\mathfrak{a}} \in I_o \cap \vec{A},\ f(\vec{\mathfrak{a}}) \in \tau,\ |\vec{\mathfrak{a}}| < x\}.$$

The smooth subsets of $\mathcal{T}$ are defined as in §I.7. If τ is a smooth subset of $\mathcal{T}$, then $C(\tau)$ denotes the smoothness constant of τ from the definition (I.7.1). Finally, μ denotes the Haar measure on $\mathcal{T}$ normalised by the condition $\mu(\mathcal{T}) = 1$.

Theorem 1. Suppose that (1) holds and let τ be a smooth subset of $\mathcal{T}$. Then

$$\pi(\tau,\vec{A};x) = \frac{\mu(\tau)\delta(\vec{A})}{|N(H_K)|}\int_2^x \frac{du}{\log u} + O(C(\tau)x\exp(-C_1(k)\sqrt{\log x})), \tag{4}$$

where $C_1(\vec{k}) > 0$ and $\delta(\vec{A}) = 1$ for $\vec{A} \in N(H_K)$, $\delta(\vec{A}) = 0$ when $\vec{A} \notin N(H_K)$;

$$\iota(\tau,\vec{A};x) = \frac{\mu(\tau)w_o(\vec{A})}{|H|}x + O(C(\tau)x^{1-C_2(\vec{k})}), \tag{5}$$

where $C_2(\vec{k}) > 0$ and $w_o(\vec{A}\vec{B}) = w_o(\vec{A})$ for $\vec{B} \in N(H_K)$. Here $C_1(\vec{k})$, $C_2(\vec{k})$ and the constants implied by the O-symbol may depend on k_j, $1 \le j \le r$, but do not depend on x and τ.

Proposition 1. If the fields $k_1,\dots,k_r$ are arithmetically independent, then

$$H = N(H_K). \tag{6}$$

Proof. See [49], theorem 1 on p. 186.

Corollary 1. Suppose that $k_1,\dots,k_r$ are arithmetically independent, then $\delta(\vec{A}) = 1$ for each $\vec{A}$ and

$$w_o(\vec{A}) = \frac{\omega(K)}{L(1,\Phi)}, \tag{7}$$

where $\omega(K)$ denotes the residue of the zeta function of K and where

$$\zeta_{k_1}(s)*\ldots*\zeta_{k_r}(s) = L(s,\Phi)^{-1}\zeta_K(s). \tag{8}$$

In particular, if $d_1 = d_2 = r = 2$ and $(D(k_1|k),D(k_2|k)) = 1$, then

$$L(1,\Phi) = L(2,\chi_o), \tag{9}$$

where $D(k_j|k)$, $j = 1,2$, denotes the relative discriminant of k_j over k and where $\chi_o = \kappa$ in notations of proposition II.5.2.

Lemma 1. Suppose that $[K:k] = d$. Then

$$\sum_{A\notin N(H_K)} \pi(\mathfrak{T},\vec{A};x) \leq d|S_o(\vec{k})|. \tag{10}$$

Remark 1. Corollary 1 shows that arithmetically independent fields are in some sense statistically independent. Namely, suppose that

$$\vec{A} = (A_1,\ldots,A_r),\ \tau = \tau_1\times\ldots\times\tau_r,\ \tau_j \subseteq \mathfrak{T}_j,\ A_j \in H_j,\ 1 \leq j \leq r$$

and that $k_1,\ldots,k_r$ are arithmetically independent, then the asymptotic probability to find $\vec{\mathfrak{a}}$ in $I_o \cap \tau \cap \vec{A}$ or to find $\vec{\mathfrak{p}}$ in $S_o \cap \tau \cap \vec{A}$ is equal, respectively, to the product of probabilities of the events:

$$\text{"}f_j(\mathfrak{a}_j) \in \tau_j,\ \mathfrak{a}_j \in A_j,\ \vec{\mathfrak{a}} \in I_o\text{"}$$

or

$$\text{"}f_j(\mathfrak{p}_j) \in \tau_j,\ \mathfrak{p}_j \in A_j,\ \vec{\mathfrak{p}} \in S_o\text{"},$$

where j ranges over the interval $1 \leq j \leq r$ and τ_j is smooth for each j.

Example 1. Let p,q_1,q_2 be three distinct rational primes and suppose that

$$p \equiv q_1 \equiv q_2 \equiv 1 \pmod 4,\ \left(\frac{q_i}{p}\right) = (-1)^i \quad \text{for } i = 1,2, \tag{11}$$

where $\left(\frac{a}{b}\right)$ denotes the Legendre symbol in $\mathbb{Z}$. Let

$$k_1 = \mathbb{Q}(\sqrt{pq_1}),\ k_2 = \mathbb{Q}(\sqrt{pq_2}),\ K = \mathbb{Q}(\sqrt{pq_1},\sqrt{pq_2}). \tag{12}$$

Let

$$p = \mathfrak{p}_i^2,\ \mathfrak{p}_i \in A_i \cap S_o(k_i),\ i = 1,2;\ \vec{A} = (A_1, A_2),$$

where $\vec{A} \in H$; let A_{oi} denote the class of principal ideals in k_i and let $\vec{A}_o = (A_{o1}, A_{o2})$. We claim that

$$w_o(\vec{A}) = \frac{w_o(\vec{A}_o)}{p}. \tag{13}$$

The proof of (13) is based on the following two lemmas.

<u>Lemma 2</u>. The following relation holds true:

$$\vec{A} \notin N(H_K) \tag{14}$$

<u>Proof</u>. See [49], theorem 2 on p. 188.

<u>Lemma 3</u>. Let k_1 and k_2 be two quadratic extensions of $\mathbb{Q}$ and let $K = k_1 \cdot k_2$ be their composite; choose $\vec{B}$ in H. If there is $\vec{\mathfrak{a}}$ such that

$$\vec{\mathfrak{a}} \in \vec{B} \cap I_o \quad \text{and} \quad \text{g.c.d.}\ (|\vec{\mathfrak{a}}|, D(k_1|\mathbb{Q}), D(k_2|\mathbb{Q})) = 1,$$

then

$$\vec{B} \in N(H_K). \tag{15}$$

<u>Proof</u>. It follows from proposition 5 in [49, p. 188] and corollary 2.6 in [49, p. 193].

<u>Proof of relation (13)</u>. In notations of the example 1, suppose that $\vec{\mathfrak{a}} \in I_o \cap \vec{A}$. Since $(D(k_1|\mathbb{Q}), D(k_2|\mathbb{Q})) = p$, it follows from (14) and (15) that

$$|\vec{\mathfrak{a}}| \equiv 0 \pmod{p}. \tag{16}$$

In view of (16), we conclude that

$$I_o \cap \vec{A} = \{\vec{\mathfrak{p}}\vec{\alpha} \mid \vec{\alpha} \in I_o \cap \vec{A}_o\}, \quad \vec{\mathfrak{p}} := (\mathfrak{p}_1, \mathfrak{p}_2). \tag{17}$$

It follows from (17) that

$$\iota(\mathfrak{T}, \vec{A}; x) = \iota(\mathfrak{T}, \vec{A}_o; \frac{x}{p}). \tag{18}$$

Relation (13) is a consequence of (18) and (5).

<u>Remark 2</u>. Example 1 shows that if $S_o(\vec{k}) \neq \emptyset$ the integral ideals having equal norms are not necessarily equidistributed over the elements of H even when (1) is satisfied.

<u>Remark 3</u>. Generalised Riemann Hypothesis allows to improve the error term in (4) and to obtain the following estimate:

$$\pi(\tau, \vec{A}; x) = \frac{\mu(\tau)\delta(\vec{A})}{|N(H_K)|} \int_2^x \frac{du}{\log u} + O(C(\tau)x^{1-C_3(d)}), \; C_3(d) > 0, \tag{R 19}$$

as soon as condition (1) is satisfied. Here the O-constant may depend on the sequence of the fields $\{k_j \mid 1 \leq j \leq r\}$ but not on x and τ.

<u>Proof of lemma 1</u>. Let $\vec{\mathfrak{p}} \in S_o$; then there is a prime divisor p in $S_o(k)$ for which $\vec{\mathfrak{p}} \mid p$. If $p \notin S_o(\vec{k})$ it follows from lemma II.5.8 that, in notations II.5.3, $\mathcal{H}(\vec{\mathfrak{p}}) \neq \emptyset$ and therefore there is a class $\vec{A}$ such that

$$\vec{\mathfrak{p}} \in \vec{A} \quad \text{and} \quad \vec{A} \in N(H_K).$$

Thus if $\vec{A} \notin N(H_K)$ we have

$$S_o \cap \vec{A} \subseteq \{\vec{\mathfrak{p}} \mid \vec{\mathfrak{p}} \in S_o, \; \vec{\mathfrak{p}} \mid p \text{ for some } p \text{ in } S_o(\vec{k})\},$$

so that

$$|S_o \cap \vec{A}| \leq d|S_o(\vec{k})|. \tag{20}$$

Inequality (10) is an immediate consequence of (20).

Lemma 4. Suppose that condition (1) is satisfied and let

$$\vec{\chi} = (\chi_1,\ldots,\chi_r),\ \chi_j \in \mathrm{gr}(k_j),\ 1 \le j \le r.$$

If

$$\prod_{j=1}^{r} \chi_j \circ N_{K/k_j} = 1, \tag{21}$$

then χ_j is a character of finite order for each j.

Proof. Let $G_j = G(k_j|\mathbb{Q})$, $1 \le j \le r$. In view of (1),

$$G \cong G_1 \times \ldots \times G_r,\quad G := G(K|\mathbb{Q}). \tag{22}$$

By (21),

$$\chi_j \circ N_{K/k_j} = \prod_{i \ne j} (\chi_i \circ N_{K/k_i})^{-1},\quad 1 \le j \le r. \tag{23}$$

Let C_K and C_j, $1 \le j \le r$, denote the idèle class groups of K and k_j, respectively. If $\alpha \in N_{K/k_j} C_K$ and $\sigma \in G_j$, then

$$\chi_j(\alpha^\sigma) = \chi_j((N_{K/k_j}\beta)^\sigma) = \chi_j\Big(\prod_{\tau \in G(K|k_j)} \beta^{\tau\sigma}\Big), \tag{24}$$

where $\alpha = N_{K/k_j}\beta$, $\beta \in C_K$. Since, by (22),

$$G(K|k_j) \cong G_1 \times \ldots \times G_{j-1} \times G_{j+1} \times \ldots \times G_r, \tag{25}$$

we have

$$\prod_{\tau \in G(K|k_j)} \beta^{\tau\sigma} = N_{K/k_j}(\beta^\sigma). \tag{26}$$

Relations (23), (24) and (26) give:

$$\chi_j(\alpha^\sigma) = \prod_{i \ne j} \chi_i(N_{K/k_i}\beta^\sigma)^{-1} \text{ for } \alpha = N_{K/k_j}\beta,\ \sigma \in G_j. \tag{27}$$

It follows from (25) that

$$N_{K/k_i}\beta^\sigma = N_{K/k_i}\beta \text{ for } \beta \in C_K,\ \sigma \in G_j,\ i \ne j. \tag{28}$$

Thus we get from (27), (28) and (23):

$$\chi_j(\alpha^\sigma) = \chi_j(\alpha) \text{ for } \alpha \in N_{K/k_j}C_K, \quad \sigma \in G_j. \tag{29}$$

Let

$$[C_K : N_{K/k_j}C_j] = \ell_j. \tag{30}$$

By (29) and (30),

$$\chi_j(\alpha)^{\ell_j} = \chi_j(\alpha^\sigma)^{\ell_j} \text{ for } \alpha \in C_j, \quad \sigma \in G_j. \tag{31}$$

Let $\chi_j^{\ell_j} = \psi_j$ and let λ_j denote the restriction of ψ_j to $C_{\mathbb{Q}}$, the idèle-class group of $\mathbb{Q}$. If $\mathfrak{p} \in S_o(k_j)$ and $N_{k_j/\mathbb{Q}}\mathfrak{p} = p^f$, then

$$\psi_j(\mathfrak{p})^{d_j} = \psi_j(p^f),$$

in view of (31). Thus

$$\psi_j^{d_j} = \lambda_j \circ N_{k_j/\mathbb{Q}}, \quad \lambda_j \in gr(\mathbb{Q}), \quad \chi_j^{\ell_j} = \psi_j, \quad 1 \leq j \leq r. \tag{32}$$

Since <u>each</u> character in $gr(\mathbb{Q})$ is of finite order, the statement of lemma 4 follows from (32).

<u>Proof of theorem 1</u>. We return to notations of §I.7 and let

$$J_1 = \vec{A} \cap I_o, \quad J_2 = S_o \setminus \tilde{S}_o(\vec{k}), \quad J_i(x) = \{\vec{\alpha} \mid \vec{\alpha} \in J_i, \ |\vec{\alpha}| < x\},$$

where $\vec{A} \in H$ and $\tilde{S}_o(\vec{k}) = \{\vec{\mathfrak{p}} \mid \vec{\mathfrak{p}} \in S_o, \ \vec{\mathfrak{p}} | p$ for some p in $S_o(\vec{k})\}$. Let $\varphi \in \hat{\mathfrak{F}}$ and let $\vec{\psi} = \varphi \circ f$; then $\vec{\psi} \in Gr$, $\vec{\psi} = (\psi_1, \ldots, \psi_r)$ with

$$\mathfrak{F}(\psi_j) = 1 \text{ for } 1 \leq j \leq r. \tag{33}$$

Thus

$$\sum_{\vec{\alpha} \in J_1(x)} \varphi(f(\vec{\alpha})) = \sum_{\substack{|\vec{\alpha}| < x \\ \vec{\alpha} \in J_1}} \vec{\psi}(\vec{\alpha}) = \sum_{\substack{|\vec{\alpha}| < x \\ \vec{\alpha} \in I_o}} \vec{\psi}(\vec{\alpha}) \frac{1}{|H|} \sum_{\vec{\lambda} \in \hat{H}} \vec{\lambda}(\vec{\alpha}) \vec{\lambda}(\vec{A}^{-1}). \tag{34}$$

In view of (1), it follows from theorem 1.1, corollary II.5.1 and corollary II.5.2 that

$$\sum_{|\vec{\alpha}| < x} \vec{\psi}(\vec{\alpha}) \vec{\lambda}(\vec{\alpha}) = w(\vec{\psi}\vec{\lambda})x + O(C_1'(\varepsilon; nd) a(\chi) b(\chi) x^{1 - \frac{2}{d+2} + \varepsilon}), \ \varepsilon > 0, \vec{\alpha} \in I_o,$$

where $\chi := \prod_{j=1}^{r} (\psi_j \circ N_{K/k_j})(\lambda_j \circ N_{K/k_j})$ and where

$w(\vec{\psi}\vec{\lambda}) = g(\chi) \dfrac{\omega(K)}{L_1(1,\vec{\psi}\vec{\lambda})}$, $\omega(K)$ being the residue of $\zeta_K(s)$ at $s=1$,

$$L_1(s,\vec{\psi}\vec{\lambda}) := (L(s,\psi_1\lambda_1)*\ldots*L(s,\psi_r\lambda_r))^{-1}L(s,\chi); \qquad (35)$$

as always,

$$g(\chi) = \begin{cases} 1, & \chi = 1 \\ 0, & \chi \neq 1 \end{cases} \quad ; \quad g(\vec{\psi}) = \begin{cases} 1, & \vec{\psi} = 1 \\ 0, & \vec{\psi} \neq 1 \end{cases} \quad \text{for } \chi \in gr(K), \vec{\psi} \in Gr.$$

Let

$$N(H_K)^{\perp} = \{\vec{\lambda} \mid \vec{\lambda} \in \hat{H},\ \vec{\lambda}(\vec{A}) = 1 \text{ for } \vec{A} \in N(H_K)\}.$$

By lemma 4, we have

$$g(\chi) = 1 \Leftrightarrow (\vec{\psi} = 1 \wedge \vec{\lambda} \in N(H_K)^{\perp}).$$

Therefore

$$\sum_{\vec{\lambda} \in \hat{H}} \vec{\lambda}(\vec{A}^{-1}) w(\vec{\psi}\vec{\lambda}) = g(\vec{\psi})\omega_K \sum_{\vec{\lambda} \in N(H_K)^{\perp}} \vec{\lambda}(\vec{A}^{-1}) L_1(1,\vec{\lambda})^{-1}. \qquad (36)$$

Estimate (5) with

$$w_o(\vec{A}) = \omega(K) \sum_{\vec{\lambda} \in N(H_K)^{\perp}} \vec{\lambda}(\vec{A}^{-1}) L_1(1,\vec{\lambda})^{-1} \qquad (37)$$

follows from (34), (36) and (I.7.14), (I.7.17), (I.7.18) in the pfoof of proposition (I.7.2). Let now

$$M = \mathfrak{T} \times N(H_K), \quad h: \vec{\mathfrak{a}} \mapsto (f(\vec{\mathfrak{a}}), \vec{A}) \qquad \text{for } \vec{\mathfrak{a}} \in I \cap \vec{A}.$$

The argument used in the proof of lemma 1 shows that

$$h(J_2) \subseteq M. \qquad (38)$$

Let $M_o = \mathfrak{T} \times H$ and let $\varphi \in \hat{M}_o$, $\vec{\psi} = \varphi \circ h$. Then $\vec{\psi} \in Gr$ and (33) is satisfied (when one writes again $\vec{\psi} = (\psi_1,\ldots,\psi_n))$. Thus

$$\sum_{\vec{\mathfrak{p}} \in J_2(x)} \varphi(h(\vec{\mathfrak{p}})) = \sum_{\substack{\vec{\mathfrak{p}} \in S_o \\ |\vec{\mathfrak{p}}|<x}} \vec{\psi}(\vec{\mathfrak{p}}) + O(d|S_o(\vec{k})|).$$

Therefore it follows from (1.6) and (1.41) that

$$\sum_{\vec{\mathfrak{p}} \in J_2(x)} \varphi(h(\vec{\mathfrak{p}})) = \sum_{\substack{p \in S_o(k) \\ |p|<x}} \chi(p) + O(d\sqrt{x}) + O(d|S_o(\vec{k})|),$$

where, in notations of proposition 1.3, we let $\chi = \mathrm{tr}\ \rho$ (cf. (1.5)). In view of (1), it follows from theorem I.5.1, corollary II.5.1 and corollary II.5.2 that

$$\sum_{\vec{\mathfrak{p}} \in J_2(x)} \varphi(h(\vec{\mathfrak{p}})) = g(\psi) \int_2^x \frac{du}{\log u} + O(x \exp(-C_4(K) \frac{\log x}{\sqrt{\log x} + \log b(\psi)})), \tag{39}$$

where $\psi \in gr(K)$, $\psi = \prod_{j=1}^{r} \psi_j \circ N_{K/k_j}$ and the O-constant may depend on the possible exceptional zero of $L(s,\psi)$ (but not on $b(\psi)$, nor on x). It follows from lemma 4 that

$$g(\psi) = 1 \Leftrightarrow \varphi_{|M} = 1. \tag{40}$$

For $\vec{A} \in N(H_K)$ relation (4) is a consequence of (39), (40) and relations (I.7.14), (I.7.17), (I.7.18). On the other hand, if $\vec{A} \in H \setminus N(H_K)$ then (4) follows from (10). This completes the proof of theorem 1.

<u>Proof of Corollary 1</u>. Suppose that the fields $k_1,\ldots,k_r$ are arithmetically independent. Then $H = N(H_K)$ by Proposition 1, and therefore relations (7) and (8) follow from (35) and (37).

§3. Equidistribution of integral points in the algebraic sets defined by a system of norm-forms.

Theorem 2.1 may be regarded, in the spirit of Appendix I.2, as an assertion about equidistribution of integral points. To simplify our considerations let us assume that

$$d_j = 2r_{2j}\ , \quad 1 \le j \le r, \tag{1}$$

where r_{2j} denotes the number of complex places of k_j. Let $\vec{A} \in H$, $\vec{A} = (A_1,\ldots,A_r)$ and let f_j be a norm-form associated to A_j by the equation (I.A.1). In view of (1), the form f_j is positive definite. We summarize the results and notations of §I.A.2 in a few commutative diagrams:

$$\begin{array}{ccc} \mathbb{R}_*^{d_j} & \xrightarrow{\ g_j\ } & X_j \\ {\scriptstyle f_j}\downarrow & \swarrow{\scriptstyle N} & \\ \mathbb{R}_+ & & \end{array} \qquad\qquad \begin{array}{ccc} \mathbb{Q}_*^{d_j} & \xrightarrow{\ \lambda_j\ } & A_j \\ {\scriptstyle g_j}\downarrow & & \downarrow{\scriptstyle \psi_j} \\ X_j^* & \xrightarrow{\ \pi_j\ } & \mathcal{T}_j \end{array}$$

$$\begin{array}{ccc} V_j^{(m)} & \xrightarrow{\ \lambda_j\ } & A_j \\ {\scriptstyle h_j}\downarrow & & \downarrow{\scriptstyle \psi_j} \\ W_j & \xrightarrow{\ \pi_j\ } & \mathcal{T}_j \end{array} \qquad\qquad h_j(a) = \frac{g_j(a)}{f_j(a)^{1/d_j}} \quad \text{for} \quad a \in \mathbb{R}_*^{d_j}$$

Here $\lambda_j, \psi_j, g_j, \pi_j$ stand for the maps denoted $\lambda_A, \psi, g_A, \pi$ in §I.A.2 when A_j and k_j take the place of A and k, respectively; $\mathcal{T}_j$ denotes the basic torus associated to k_j as in §I.1, X_j and W_j are defined as X and W in §I.A.2, finally

$$V_j^{(t)} = \{a \mid a \in \mathbb{R}^{d_j},\ \ f_j(a) = t\}, \quad t \ge 0.$$

We write, for brevity,

$$\mathbb{Q}_*^{d_j} := \mathbb{Q}^{d_j}\ \{0\}\ , \quad \mathbb{R}_*^{d_j} = \mathbb{R}^{d_j} \setminus V_j^{(0)}$$

For $U_j \subseteq W_j$, $m \in \mathbb{Z}$, $m > 0$, let

$$B_j(m,U_j) = \{\vec{a} \mid \vec{a} \in \mathbb{Z}^{d_j},\ h_j(\vec{a}) \in U_j,\ f_j(\vec{a}) = m\}$$

and let

$$\mathfrak{b}_j(m,U_j) = \{\mathfrak{a} \mid \mathfrak{a} \in I_o(k_j) \cap A_j,\ \psi_j(\mathfrak{a}) \in \pi_j(U_j),\ |\mathfrak{a}| = m\}.$$

Write

$$W = W_1 \times \ldots \times W_r.$$

Definition 1. Let $U \subseteq W$. We say that U is toroidal if

$$U = U_1 \times \ldots \times U_r\ ,\quad U_j \subseteq W_j \quad \text{for} \quad 1 \leq j \leq r\ ,$$

and U_j is toroidal, for each j, in the sense of §I.A.2.

For $U = U_1 \times \ldots \times U_r$ with $U_j \subseteq W_j$, $1 \leq j \leq r$, let

$$B(m,U) = B_1(m,U_1) \times \ldots \times B_r(m,U_r)$$

and let

$$\mathfrak{b}(m,U) = \mathfrak{b}_1(m,U_1) \times \ldots \times \mathfrak{b}_r(m,U_r).$$

Furthermore, let

$$b(U,x) := \text{card} \bigcup_{1 \leq m < x} B(m,U).$$

Write

$$\mathcal{T} = \mathcal{T}_1 \times \ldots \times \mathcal{T}_r,\quad \pi\colon W \to \mathcal{T},\quad \pi = \pi_1 \times \ldots \times \pi_r.$$

Proposition 1. If U is a toroidal subset of W, then

$$b(U,x) = \iota(\pi(U),\vec{A},x), \tag{2}$$

in notations of §2.

Proof. By definition,

$$\iota(\pi(U),\vec{A},x) = \mathrm{card}(\bigcup_{1\leq m<x} \mathfrak{b}(m,U)). \tag{3}$$

On the other hand, since U is toroidal it follows that λ defines a one-to-one correspondence between $B(m,U)$ and $\mathfrak{b}(m,U)$ for each m (such that $m \geq 1$, $m \in \mathbb{Z}$). Therefore (3) implies (2).

<u>Notations 1</u>. Let

$$V = \{\vec{a}\,|\,\vec{a} = (a_1,\dots,a_r),\ a_j \in \mathbb{R}^{d_j} \text{ for } 1 \leq j \leq r,$$

$$f_1(a_1) = \dots = f_r(a_r)\},$$

$$V(\mathbb{Z}) = V \cap (\mathbb{Z}^{d_1}\times\dots\times\mathbb{Z}^{d_r}),$$

$$V^{(m)} = V_1^{(m)}\times\dots\times V_r^{(m)} \quad \text{for } m \in \mathbb{N}.$$

Let, for $U \subseteq W$, $x > 0$,

$$b(U,x) = \{\vec{a}\,|\,\vec{a} \in V(\mathbb{Z}),\ h(\vec{a}) \in U,\ |\vec{a}| < x\},$$

where

$$h\colon \mathbb{R}_*^{d_1}\times\dots\times\mathbb{R}_*^{d_r} \to W, \quad h = h_1\times\dots\times h_r,$$

and where

$$|\vec{a}| := \frac{1}{r}\sum_{j=1}^{r} f_j(a_j) \quad \text{for } \vec{a} = (a_1,\dots,a_r),\ a_j \in \mathbb{R}^{d_j}.$$

<u>Remark 1</u>. Clearly, if $U = U_1\times\dots\times U_r$, $U_j \subseteq W_j$ for $1 \leq j \leq r$, then

$$\{\vec{a}\,|\,\vec{a} \in V(\mathbb{Z}),\ h(\vec{a}) \in U,\ |\vec{a}| < x\} = \bigcup_{1\leq m<x} B(m,U).$$

Therefore the two definitions of $b(U,x)$ agree.

<u>Notations 2</u>. Let μ_j be the Haar measure on $\mathfrak{T}_j$ normalised by the condition $\mu_j(\mathfrak{T}_j) = 1$ and let $\tilde{\mu}_j$ be the corresponding v_j^*-invariant measure on W_j defined as in §I.A.2. We write then

$$\mu = \mu_1\times\dots\times\mu_r, \quad \tilde{\mu} = \tilde{\mu}_1\times\dots\times\tilde{\mu}_r$$

for the product measures on $\mathcal{T}$ and W, respectively.

Definition 2. Let $U \subseteq W$. We say that U is rectangular if U is toroidal and $\pi(U)$ is an elementary subset of $\mathcal{T}$ in the sense of (I.7.7). Let

$$E = \{U | U \subseteq W, \quad U \text{ is rectangular}\}.$$

Proposition 2. Suppose that the fields $k_1, \dots, k_r$ are arithmetically independent and that condition (1) is satisfied. Then

$$b(U,x) = \tilde{\mu}(U) \frac{\omega(K)x}{L(1,\Phi)|H|} + O(C(U)x^{1-C_1(\vec{k})}), \; C_1(\vec{k}) > 0, \qquad (4)$$

for any $(E,\tilde{\mu})$-smooth subset U of W. Here $C(U)$ is the smoothness constant of U; the implied O-symbol constant and $C_1(\vec{k})$ may depend on k_j, $1 \leq j \leq r$, but not on x and U.

Proof. Relation (4) follows from (2.5), (2.7) and (2) in view of proposition 2.1 and proposition I.7.1.

One remarks that

$$V = V^{(0)} \cup (V^{(1)} \times \mathbb{R}_+).$$

Let λ be the restriction to $\mathbb{R}_+$ of the Lebesque measure on $\mathbb{R}$, we define a measure μ' on V by letting

$$\mu'(V^{(0)}) = 0, \; \mu'(V^{(1)} \times (0,1]) = 0,$$

$$\mu' = (\tilde{\mu} \circ h) \times \lambda \qquad \text{on} \quad V^{(1)} \times (\mathbb{R}_+ \setminus (0,1]).$$

Let

$$E_1 = \{U \times (t_1,t_2] \,|\, U \subseteq V^{(1)}, \; h(U) \in E, \; 0 < t_1 < t_2\},$$

where

$$(t_1,t_2] := \{t | t_1 < t \leq t_2\}.$$

Theorem 1. Suppose that the fields $k_1,\dots,k_r$ are arithmetically independent and condition (1) is satisfied. Let U be a (E_1,μ')-smooth subset of V and let

$$t(U) := \sup\{s \mid V^{(s)} \cap U \neq 0\}.$$

Then

$$\operatorname{card}(U\cap V(\mathbb{Z})) = \mu'(U)\,\frac{\omega(K)}{|H|\,L(1,\Phi)} + O(C(U)\,t(U)^{1-C_2(\vec{k})}),\quad {}_2(\vec{k}) > 0, \tag{5}$$

where $C_2(\vec{k})$ and the O-constants depend on k_j, $1 \leq j \leq r$, but not on U; here $C(U)$ denotes the smoothness constant of U.

Proof. Since f_j, $1 \leq j \leq r$, is positive definite on $\mathbb{Q}^{d_j}$, the subset $V^{(o)}$ contains no integral points except the origin. Let $U \in E_1$, then

$$U = U_1\times(t_1,t_2] \text{ with } 0 < t_1 < t_2,\ U_1 \subseteq V^{(1)},\ h(U_1) \in E. \tag{6}$$

It follows from (6) and (4) that

$$\operatorname{card}(U\cap V(\mathbb{Z})) = \tilde{\mu}(h(U_1))(t_2-t_1)\,\frac{\omega(K)}{|H|\,L(1,\Phi)} + O(t_2^{1-C_1(\vec{k})}).$$

Therefore (5) holds for elementary sets (i.e. for the elements of E_1). The argument used in the proof of proposition I.7.1 allows to deduce estimate (5) for any (E_1,μ')-smooth set.

Notations 3. Let, as in §I.A2,

$$|a| = \max_{1\leq j\leq \ell} |a_j| \quad \text{for } a = (a_1,\dots,a_\ell),\ a \in \mathbb{C}^\ell,$$

and let

$$U_j(x) = \{a \mid a \in \mathbb{R}^{d_j},\ |g_j(a)| < x^{\delta_j}\},\quad 1 \leq j \leq r,$$

for $x > 1$, $\delta_j := \frac{1}{d_j}$. Further, let

$$U(x) = U_1(x)\times\dots\times U_r(x);\quad U'(x) = U(x) \cap V.$$

Lemma 1. The following estimate holds:

$$C(U'(x)) = O(\log x)^{\nu(\vec{k})}), \quad \nu(\vec{k}) \geq 0, \tag{7}$$

where $\nu(\vec{k})$ and the O-constant may depend on k_j, $1 \leq j \leq r$, but not on x.

Proof. Let

$$\{\varepsilon_{ji} \mid 1 \leq i \leq r_{2j}-1\}$$

be a system of fundamental units in k_j, $1 \leq j \leq r$. Suppose that

$$\alpha \in h_j(V_j^{(1)} \cap U_j(x)) \quad \text{and} \quad \sigma(\varepsilon)\alpha \in h_j(V_j^{(1)} \cap U_j(x)) \tag{8}$$

with

$$\varepsilon = \prod_i \varepsilon_{ji}^{n_i}, \quad n_i \in \mathbb{Z}, \quad 1 \leq i \leq r_{2j}-1, \tag{9}$$

where $\sigma\colon k_j \to X_j$ denotes the diagonal embedding of k_j in X_j. By definition of $U_j(x)$, it follows from (8) and (9) that

$$\log|\alpha_p| + \sum_{1\leq i\leq r_{2j}-1} n_i \log|\varepsilon_{ji}^{(p)}| \leq \delta_j \log x, \quad p \in S_\infty(k_j), \tag{10}$$

where $\varepsilon_{ji}^{(p)}$ and α_p denote the projections of $\sigma(\varepsilon_{ji})$ and α on k_{jp}, respectively. Inequality (10) combined with the equations

$$\sum_{p\in S_\infty(k_j)} \log|\alpha_p| = \sum_{p\in S_\infty(k_j)} \log|\varepsilon_{ji}^{(p)}| = 0 \tag{11}$$

leads to an estimate for n_i:

$$n_i = O(\log x), \quad 1 \leq i \leq r_{2j}-1.$$

Since

$$V^{(1)} \cap U(x) = V_1^{(1)} \cap U_1(x) \times \ldots \times V_r^{(1)} \cap U_r(x)$$

it follows from (11) that there is a covering

$$V^{(1)} \cap U(x) \subseteq \bigcup_{1 \le p \le N} \kappa_p\,, \quad h(\kappa_p) \in E, \tag{12}$$

with

$$N = O((\log x)^a), \qquad a := \sum_{j=1}^{r} (r_{2j}-1). \tag{13}$$

Relations (12) and (13) imply (7) by the definition of the smoothness constant.

<u>Lemma 2.</u> There is $C_3(\vec{k})$ such that

$$\mu'(U'(x)) = C_3(\vec{k})x + O_\varepsilon(x^\varepsilon), \quad C_3(\vec{k}) > 0, \quad \varepsilon > 0, \tag{14}$$

where $C_3(\vec{k})$ and the O_ε-constant may depend on k_j but not on x.

<u>Proof</u>. By definition,

$$\mu'(U'(x)) = \int_1^x dt\ \tilde{\mu}(h(V^{(t)} \cap U(x))), \tag{15}$$

and

$$h(V^{(t)} \cap U(x)) = h_1(V_1^{(1)} \cap U_1(\tfrac{x}{t})) \times \ldots \times h_r(V_r^{(1)} \cap U_r(\tfrac{x}{t})),$$

so that

$$\tilde{\mu}(h(V^{(t)} \cap U(x))) = \prod_{j=1}^{r} \tilde{\mu}_j(h_j(V_j^{(1)} \cap U_j(\tfrac{x}{t}))). \tag{16}$$

On the other hand, it follows from lemma I.A2.1 that

$$\tilde{\mu}_j(h_j(V_j^{(1)} \cap U_j(\tfrac{x}{t}))) = C_4(k_j)(\log \tfrac{x}{t})^{r_{2j}-1}, \quad 1 \le j \le r, \tag{17}$$

with $C_4(k_j) > 0$. Relation (14) is a consequence of (15) - (17).

<u>Theorem 2</u>. Suppose that the fields k_j, $1 \le j \le r$, are arithmetically independent and condition (1) is satisfied. Then there are $C_5(\vec{k})$ and $C_6(\vec{k})$ depending on k_j only and such that

$$\mathrm{card}(U(x) \cap V(\mathbb{Z})) = C_5(\vec{k})x + O(x^{1-C_6(\vec{k})}), \ C_5(\vec{k}) > 0, \ C_6(\vec{k}) > 0, \tag{18}$$

with an O-constant independent of x.

Proof. Since $t(U'(x)) = O(x)$, it follows from theorem 1 and lemma 1 that

$$\operatorname{card}(U'(x) \cap V(\mathbb{Z})) = \mu'(U'(x)) \frac{\omega(K)}{|H| L(1,\Phi)} + O(x^{1-C_6(\vec{k})}) \tag{19}$$

with $C_6(\vec{k}) > 0$. Relation (18) is a consequence of (19) and lemma 2.

Remark 2. The same argument leads to an estimate

$$\operatorname{card}(\tilde{U}(x) \cap V(\mathbb{Z})) = \mu'(\tilde{U}(x)) \frac{\omega(K)}{|H| L(1,\Phi)} + O(x^{1-C_7(\vec{k})}) \tag{20}$$

with $C_7(k) > 0$ when one takes

$$\tilde{U}(x) = \{\vec{a} \mid \vec{a} \in V,\ \vec{a} = (a_1, \ldots, a_r), |a_j| \in \mathbb{R}^{d_j},\ |a_j| < x^{\delta_j},\ 1 \le j \le r\}.$$

Chapter IV. Remarks and comments.

Our goal in this chapter is twofold: we relate here the contents of this book to the work of other authors and acknowledge our indebtedness to these authors.

Chapter I. Although the methods used here are of classical origin, some of the results are new while the other ones are not easily accessible in the literature. Our exposition presupposes acquaintance with the class field theory as it is treated, for example, in [22],[93],[69],[17],[70].

§1. The best references concerning the material presented here and known to us are the original paper of E. Hecke, [24], a paragraph from H. Hasse, [23], and the books [25], [93] (cf. also the beginning of [62]).

§2. We follow mostly [48] generalising the results we need to the case of compact groups. Alternatively the reader may consult well known monographs, [92],[77],[83] (cf. also [51] for results related to proposition 2).

§3. Here we try to explain to ourselves and to the reader some of the things we have learned from the articles [91],[7],[47],[87],[26] (cf. also [94] for a discussion of Artin-Weil conjectures and Riemann hypothesis).

§4-6. Making use of the technique developed by E. Landau, [40]-[43], and of the convexity theorems proved in [80] we obtain estimates for character sums extended over prime ideals or over integral ideals with effective numerical constants in the error terms. Although such estimates have been discussed by several authors, [39],[74],[84], some of the results described here are new (cf. [68]). Lemma 6.1 is a generalisation of a classical argument due to J.E. Littlewood (see, e.g., [89], Chapter XIV). We should like to quote the following articles for related material and further references: [32],[52],[13],[3],[4],[45],[28],[75], [95],[53].

§7. We follow [62],[65] (cf. [79],[32],[52] for studies in the multidimensional arithmetic of E. Hecke, [24]). Concerning equidistribution problems one may consult [96],[33],[85, Appendix to Ch. 1] and [46, Ch. VI, §2].

Appendix I. Although we have not found this result in the literature, it is probably known to many authors.

Appendix II. We follow [65]. The relation between ideal classes and norm-forms is known from the classical literature (cf., for instance [25],[2]).

Chapter II. This chapter is intended as a detailed exposition of the results summarised in [63] and appearing elsewhere, [66]. It serves as an illustration of the methods described in Chapter I.

§1. Convolution of power series is a classical notion, [21] (cf. also [88, §4.6]). Scàlar products of Dirichlet series have been widely discussed in the context of automorphic functions (cf., for example, [81],[76],[73],[29],[30],[19],[35] and references therein).

§§2-4. The construction presented here is a generalisation of the one described in [34] to compact groups (cf. [35]-[38],[60],[61],[64],[63], [66]). We refer to [44],[12],[6] for some results concerning Euler products defined over $\mathbb{Q}$ and having analogous properties (cf. [61] for a review of these results).

§5. The scalar product of L-functions "mit Größencharakteren", a concept due to Yu. V. Linnik, [50] (cf. also [54],[14]) has been studied by several authors. After some work done by the author of this book and related to the case of two quadratic fields, [54]-[57], A.I. Vinogradov, [90], continued $L(s,\vec{\psi})$ to the half-plane $\operatorname{Re} s > \frac{1}{2}$. O.M. Fomenko,[14], continued $L(s,\vec{\psi})$ to the whole plane in the case of two quadratic extensions of $\mathbb{Q}$, while H. Jacquet, [29], proved an analogous result for two quadratic extensions of any number field. P.K.J. Draxl, [9], continued $L(s,\vec{\psi})$ meromorphically to $\mathbb{C}_+$ without any a priori assumptions about k_j, $1 \leq j \leq r$. Several authors, [59],[15],[36], independently and approximately at the same time noticed that in the case of two quadratic fields $L(s,\vec{\psi})$ can be explicitly evaluated. N. Kurokawa, [34], proved theorem 1 assuming ρ_j, $1 \leq j \leq r$, in (II.2.6) is of Galois type and ψ_j, $1 \leq j \leq r$, in (II.5.3) is of finite order. We have removed this assumption: at first assuming the (Generalised) Riemann Hypothesis and considering only $L(s,\vec{\psi})$ (cf. [60],[61]), then proving theorem II.5.1 in full generality, [64],[63],[66]. Recently N. Kurokawa, [37],[38], has also announced results implying theorem III.5.1. Lemma 3 goes back to [34]; lemma 8 is a generalisation of the results described in [58] (cf. [49]).

§6. We follow [66] (cf. also [64], §III.2, where the idea of the proof has been sketched). Conjecture 1 can be deduced from a uniform estimate of the volume of a tube around a real analytic (or even semialgebraic) set; we have not found a complete proof of the required estimate in the published literature (cf., however, [84, p. 145],[98],[99] where such an estimate has been claimed or conjectured). We refer in this connection to the known results about volumes of tubes around smooth manifolds, [97],[27],[18]. The author is indebted to Professor S. Markvorsen for these references and discussions of this geometric problem and to Professor Y. Yomdin for the information about his results described in [98].

Chapter III. It is an alternative exposition of the results appearing in [62],[49],[65],[68].

§1. The estimates for the moments of the coefficients of L-functions Hecke have been discussed by several authors (cf. [90],[5],[62],[72], [82] and references therein). Theorem 1 has been proved in [68] (cf. [67] and [64] for a slightly less precise version of this result); we refer to [43], §242, Satz 62 for the argument similar to the one used in the proof of lemma 4 and lemma 5 and to [31] for a discussion of a problem related to the estimate obtained here. The author is indebted to Professor H. Delange for an alternative proof of lemma 5 and to Professor E.-U. Gekeler for pointing out that estimate (III.1.39) would imply the Artin-Weil conjecture.

§2. Here we follow mostly [62],[49]. The idea of example 1 is due to P. Deligne, [8]. We refer to [90],[10],[11],[71] for some results related to the problems discussed here.

§3. We follow [65]; theorem 2 will appear in [100].
We refer to [86] for the analytic results about distribution of integral points in algebraic sets; representation of integers by a norm-form has been discussed by several authors (cf. [20] and references therein). The author is grateful to Professor P. Deligne and Professor O. Gabber for several important conversations related to the problems discussed in this chapter.

Literature cited.

[1] M. Abramowitz, I.A. Stegun, Handbook of Mathematical functions, Dover Publications, New York, 1972.

[2] Z.I. Borevich, I.R. Schafarevich, Number Theory, Academic Press, New York-London, 1966.

[3] K. Chandrasekharan, R. Narasimhan, Functional equations with multiple gamma factors and the average order of arithmetical functions, Annals of Mathematics, 76 (1962), p. 93 - 136.

[4] K. Chandrasekharan, R. Narasimhan, The approximate functional equation for a class of zeta-functions, Mathematische Annalen, 152 (1963), p. 30 - 64.

[5] K. Chandrasekharan, A. Good, On the number of integral ideals in Galois extensions, Monatshefte für Mathematik, 95 (1983), p. 99-109.

[6] G. Dahlquist, On the analytic continuation of Eulerian products, Arkiv för Mathematik, 1 nr 36 (1952), p. 533 - 554.

[7] P. Deligne, Les constantes des équations fonctionnelles des fonctions L, Springer Lecture Notes in Mathematics, 349 (1973), p. 501 - 595.

[8] P. Deligne, Private communication, 1982.

[9] P.K.J. Draxl, L-Funktionen Algebraischer Tori, Journal of Number Theory, 3 (1971), no. 4, p. 444 - 467.

[10] P.K.J. Draxl, Fonctions L et représentation simultanée d'un nombre premier par plusieurs formes quadratiques, Séminaire Delange-Pisot-Poitou, 12ème année 1970/71, Paris, No. 12.

[11] P.K.J. Draxl, Remarques sur le groupe de classes du composé de deux corps de nombres linéairement disjoints, Séminaire Delange-Pisot-Poitou, 12ème année 1970/71, Paris, No. 24.

[12] T. Estermann, On certain functions represented by Dirichlet series, Proceedings of the London Mathematical Society, 27 (1928), p. 435 - 448.

[13] E. Fogels, On the zeros of Hecke's L-functions I, Acta Arithmetica 7 (1961/62), p. 87 - 106.

[14] O.M. Fomenko, Analytic continuation to the whole plane and the functional equation of the scalar product of Hecke L-functions for two quadratic fields (in Russian), Proceedings of the Steklov Mathematical Institute, 128 part II (1972), p. 232 - 241.

[15] E. Gaigalas, The distribution of primes in two quadratic imaginary number fields. I, II (in Russian), Litovskij Mathematical Sbornik, 19 (1979), no. 2, p. 45 - 60 and no. 4, p. 69 - 75.

[16] E. Gaigalas, The scalar product of Hecke L-series of quadratic fields (in Russian), Litovskij Mathematical Sbornik, 15 (1975), no. 4, p. 41 - 52.

[17] L.J. Goldstein, Analytic number theory, Prentice Hall, Englewood Cliffs, N.J., 1971.

[18] A. Gray, L. Vanhecke, The volumes of tubes in a Riemannian manifold, Rend. Sem. Mat. Univ. Polytechn. Torino, 39 (1983), no. 3, p. 1 - 50 (MR 84i:53053).

[19] B. Gross, D. Zagier, Points de Heegner et dérivées de fonctions L, Comptes Rendus de l'Académie des Sciences (Paris), 297 (1983), Série I (19 Septembre), p. 85 - 87.

[20] K. Györy, A. Pethö, Über die Verteilung der Lösungen von Normformen Gleichungen, Acta Arithmetica, 37 (1980), p. 143 - 165.

[21] J. Hadamard, Théorème sur les séries entières, Acta Mathematica, 22 (1899), p. 55 - 63.

[22] H. Hasse, Zahlbericht, Physica Verlag, Würzburg-Wien, 1965.

[23] H. Hasse, Zetafunktionen und L-functionen zu Funktionenkörpern von Fermatschem Typus, Abhandlungen Deutscher Akademie der Wissenschaften, Berlin, Kl. Mathem.-Nat. 4 (1954).

[24] E. Hecke, Eine neue Art von Zetafunktionen und ihre Beziehungen zur Verteilung der Primzahlen (Zweite Mitteilung), Mathematische Zeitschrift, 6 (1920), p. 11 - 51.

[25] E. Hecke, Vorlesungen über die Theorie der algebraischen Zahlen, Chelsea Publishing Company, New York, 1970.

[26] H. Heilbronn, Zeta-functions and L-functions, in: Algebraic number theory (edited by J.W.S. Cassels and A. Fröhlich), Academic Press, 1967, p. 204 - 230.

[27] E. Heintze, H. Karcher, A general comparison theorem with applications to volume estimates for submanifolds, Annales Scientifiques de l'école normale supérieure, 11 (1978), p. 451 - 470.

[28] J. Hinz, Eine Erweiterung des nullstellenfreien Bereiches des Heckeschen Zetafunktionen und Primideale in Idealklassen, Acta Arithmetica, 38 (1980), p. 209 - 254.

[29] H. Jacquet, Automorphic forms on GL(2), Springer Lecture Notes in Mathematics, 278 (1972).

[30] H. Jacquet, Dirichlet series for the group GL(N), in: Automorphic forms (Bombay Colloquium 1979), Springer-Verlag, 1981, p. 155 - 163.

[31] J. Kaczorowski, Some remarks on factorization in algebraic number fields, Acta Arithmetica, 43 (1983), p. 53 - 68.

[32] I.P. Kubilus, On some problems in geometry of prime numbers (in Russian), Mathematical Sbornik USSR, 31 (1952), No. 3, p. 507 - 542.

[33] L. Kuipers, H. Niederreiter, Uniform distribution of sequences, Wiley-Interscience, 1974.

[34] N. Kurokawa, On the meromorphy of Euler products, Part I: Artin type, Tokyo Institute of Technology Preprint, 1977.

[35] N. Kurokawa, On the meromorphy of Euler products, Proceedings of Japan Academy, 54A (1978), p. 163 - 166.

[36] N. Kurokawa, On Linnik's problem, Proceedings of Japan Academy, 54A (1978), p. 167 - 169.

[37] N. Kurokawa, On some Euler products. I, Proceedings of Japan Academy, 60 (1984), Ser. A, p. 335 - 338.

[38] N. Kurokawa, On some Euler products. II, Proceedings of Japan Academy, 60 (1984), Ser. A., p. 365 - 368.

[39] J.C. Lagarias, H.L. Montgomery, A.M. Odlyzko, A bound for the least prime ideal in the Chebotarev density theorem, Inventiones Mathematicae, 54 (1979), p. 271 - 296.

[40] E. Landau, Über Ideale und Primideale in Idealklassen, Mathematische Zeitschrift, 2 (1918), p. 52 - 154.

[41] E. Landau, Einführung in die elementare und analytische Theorie der algebraischen Zahlen und Idealen, B.G. Teubner Verlag, 1918.

[42] E. Landau, Vorlesungen über Zahlentheorie, S. Hirzel Verlag, Leipzig, 1927.

[43] E. Landau, Handbuch der Lehre von der Verteilung der Primzahlen, Leipzig: Teubner, 1909.

[44] E. Landau, A. Walfisz, Über die Nichtfortsetzbarkeit einiger durch Dirichletsche Reihen definierter Funktionen, Rend. di Palermo, 44 (1919), p. 82 - 86.

[45] S. Lang, On the zeta function of number fields, Inventiones Mathematicae, 12 (1971), p. 337 - 345.

[46] S. Lang, Algebraic Number Theory, Addison-Wesley, 1970.

[47] R.P. Langlands, On Artin's L-functions, Rice University Studies, 56 (1970), p. 23 - 28.

[48] W. Ledermann, Introduction to group characters, Cambridge, 1977.

[49] W.-Ch. W. Li, B.Z. Moroz, On ideal classes of number fields containing integral ideals of equal norms, Journal of Number Theory, 21 (1985), No. 2, p. 185 - 203.

[50] Yu. V. Linnik, Private communication, 1962.

[51] G.W. Mackey, Induced representations of locally compact groups. I, Annals of Mathematics, 55 (1952), p. 101 - 139.

[52] T. Mitsui, Generalised prime number theorem, Japanese Journal of Mathematics, 26 (1956), p. 1 - 42.

[53] C.J. Moreno, Analytic proof of the strong multiplicity one theorem, American Journal of Mathematics, 107 (1985), p. 163 - 206.

[54] B.Z. Moroz, Analytic continuation of the scalar product of Hecke L-series for two quadratic fields and its application (in Russian), Doklady of the Academy of Sciences USSR, ser. Mathematika, 150 (1963), No. 4, p. 752 - 754.

[55] B.Z. Moroz, On analytic continuation of the scalar product of Hecke L-series for two quadratic fields (in Russian), Doklady of the Academy of Sciences USSR, ser. Mathematika, 155 (1964), No. 6, p. 1265 - 1267.

[56] B.Z. Moroz, Composition of binary quadratic forms and scalar product of Hecke series (in Russian), Proceedings of the Steklov Mathematical Institute, 80 (1965), p. 102 - 109.

[57] B.Z. Moroz, On distribution of pairs of prime divisors in two quadratic fields, I, II (in Russian), Vestnik of the Leningrad University, Ser. Mathematika, No. 19 (1965), issue 4, p. 47 - 57 and No. 1 (1966), issue 1, p. 64 - 79.

[58] B.Z. Moroz, On zeta-functions of algebraic number fields (in Russian), Mathematical Notes of the USSR Academy of Sciences, 4 (1968), No. 3, p. 333 - 339.

[59] B.Z. Moroz, On the convolution of Hecke L-functions, Mathematika, 27 (1980), p. 312 - 320.

[60] B.Z. Moroz, Scalar product of L-functions with Grössencharacters: its meromorphic continuation and natural boundary, Journal für die reine und angewandte Mathematik, 332 (1982), p. 99 - 117.

[61] B.Z. Moroz, Euler products (variation on a theme of Kurokawa's), Astêrisque, 94 (1982), p. 143 - 151.

[62] B.Z. Moroz, On the distribution of integral and prime divisors with equal norms, Annales de l'Institut Fourier (Grenoble), 34 (1984), fasc. 4, p. 1 - 17.

[63] B.Z. Moroz, Produits eulériens sur les corps de nombres, Comptes Rendus de l'Acadêmie des Sciences (Paris), 301 (1985), Série I, no. 10, p. 459 - 462.

[64] B.Z. Moroz, Vistas in analytic number theory, Bonner Mathematische Schriften, 156 (1984), Bonn.

[65] B.Z. Moroz, Integral points on norm-form varieties, Journal of Number Theory, to appear.

[66] B.Z. Moroz, On analytic continuation of Euler products, M.P.I. für Mathematik Preprint, 85 - 7, Bonn, 1985.

[67] B.Z. Moroz, On the coefficients of Artin-Weil L-functions, M.P.I. für Mathematik Preprint, 84 - 12, Bonn, 1984.

[68] B.Z. Moroz, Estimates for character sums in number fields, M.P.I. für Mathematik Preprint, 86 - 14, Bonn, 1986.

[69] J. Neukirch, Klassenkörpertheorie, Mannheim, 1969.

[70] J. Neukirch, Class field theory, Springer-Verlag, 1986.

[71] R.W.K. Odoni, A new equidistribution property of norms of ideals in given classes, Acta Arithmetica, 33 (1977), p. 53 - 63.

[72] R.W.K. Odoni, Scalar products of certain Hecke L-series and moments of weighted norm-counting functions, Canadian Mathematical Bulletin, 28 (1985), p. 272 - 279.

[73] A. Ogg, On a convolution of L-series, Inventiones Mathematicae, 7 (1969), p. 297 - 312.

[74] J. Oesterlé, Versions effectives du théorème de Chebotarev sous l'hypothèse de Riemann généralisée, Astérisque, 61 (1979), p. 165 - 167.

[75] A. Perelli, General L-functions, Annali die Matematica pura ed applicata, 80 (1982), p. 287 - 306.

[76] H. Petersson, Über die Berechnung der Skalarprodukte ganzer Modulformen, Commentarii Mathematici Helvetici, 22 (1949), p. 168 - 199.

[77] L. Pontrjagin, Topological groups, Princeton, 1946.

[78] K. Prachar, Primzahlverteilung, Springer-Verlag, 1978.

[79] H. Rademacher, Primzahlen reell-quadratischer Zahlkörpern in Winkelräumen, Mathematische Annalen, 111 (1935), p. 209 - 228.

[80] H. Rademacher, On the Phragmén-Lindelöf theorem and some applications, Mathematische Zeitschrift, 72 (1959), p. 192 - 204.

[81] R.A. Rankin, Contributions to the theory of Ramanujan's function $\tau(n)$ and similar arithmetical functions. II: The order of the Fourier coefficients of integral modular forms, Proceedings of the Cambridge Philosophical Society, 35 (1939), p. 357 - 372.

[82] R.A. Rankin, A family of new forms, Annales Academia Scientiarum Fennica, Ser. A.I. Mathematica, 10 (1985), p. 461 - 467.

[83] J -P. Serre, Représentations linéaires des groupes finis, Hermann, Paris, 1978.

[84] J -P. Serre, Quelques applications du théorème de densité de Chebotarev, Publications Mathématiques I.H.E.S., 54 (1981), p. 123 - 202.

[85] J -P. Serre, Abelian ℓ-adic representations and elliptic curves, W. Benjamin, 1968.

[86] W.M. Schmidt, The density of integer points on homogeneous varieties, Acta Mathematica, 154 (1985), p. 243 - 296.

[87] J. Tate, Number theoretic background, Proceedings of Symposia in Pure Mathematics (American Mathematical Society), 33 (1979), Part II, p. 3 - 26.

[88] E.C. Titchmarch, The theory of functions, Oxford University Press, 1979.

[89] E.C. Titchmarch, The theory of Riemann zeta-function, Oxford, 1951.

[90] A.I. Vinogradov, On the continuation to the left half-plane of the scalar product of Hecke L-series "mit Grössencharakteren" (in Russian), Izvestia of the USSR Academy of Sciences, Seria Mathematika, 29 (1965), p. 485 - 492.

[91] A. Weil, Sur la Théorie du Corps de Classes, Journal of the Mathematical Society of Japan, 3 (1951), p. 1 - 35.

[92] A. Weil, L'intégration dans les groupes topologiques et ses applications, Hermann, Paris, 1951.

[93] A. Weil, Basic number theory, Springer-Verlag, 1973.

[94] A. Weil, Sur les formules explicites de la théorie des nombres, Izvestia of the USSR Academy of Sciences, Seria Mathematika, 36 (1972), p. 3 - 18.

[95] A. Weiss, The least prime ideal, Journal für die reine und angewandte Mathematik, 338 (1983), p. 59 - 94.

[96] H. Weyl, Über die Gleichverteilung von Zahlen mod Eins, Mathematische Annalen, 77 (1916), p. 313 - 352.

[97] H. Weyl, On the volumes of tubes, American Journal of Mathematics, 61 (1939), p. 461 - 472.

[98] Y. Yomdin, Metric properties of semialgebraic sets and mappings and their applications in smooth analysis, Ben-Gurion University of Negev, Preprint, 1985.

[99] S. Smale, Lecture at the M.P.I. for Mathematics (Bonn), July 7, 1983 (unpublished).

[100] B.Z. Moroz, A footnote to my recent paper (submitted for publication).

Index